THE MEASUREMENT MANUAL

how to measure child performance

MARILYN A. COHEN, PhD
PAMELA J. GROSS, MEd

SPECIAL CHILD PUBLICATIONS / SEATTLE

Special Child Publications
J. B. Preston, Editor & Publisher
P.O. Box 33548
Seattle, Washington 98133

Serving the special child since 1962

International Standard Book Number: 0-87562-077-9

94	93	92	91	90	89	88	87	86	85
10	9	8	7	6	5	4	3	2	1

Contents

   A Definition of Behavior ★ Behavior Is Observable: Seeing Is Believing ★ Behavior Involves Action or Movement ★ *Using a Behavioral Ruler* ★ Behavior Chunks ★ Movement Cycles ★ *Conclusion*

   Getting Started: Initial Guidelines ★ Make a Shopping List ★ Decide on a Shopping Strategy ★ Shop Comparatively and Competitively ★ Avoid Impulse Purchases ★ Establish Pinpointing Priorities ★ *Exploring Pinpointing Sources and Strategies* ★ Brainstorming: An Informal Framework ★ Expert Opinion ★ Requirements Dictated by Administrative Guidelines ★ Instructional Materials, Textbook and Workbook Units ★ Formal and Informal Evaluative Tools ★ Skill Sequences and Checklists ★ *Choosing Pinpoints that Best Describe Performance* ★ Tool Behaviors ★ Describing Complex Performance Domains ★ Fair Pairs ★ Response Modes ★ Reflecting Learning Across a Wide Range of Developmental or Subject Areas ★ Accounting for All Stages of Learning ★ *Refining Behavioral Pinpoints* ★ Establishing Uniform Pinpoints ★ Functional Definitions ★ Behavior Slicing ★ *Conclusion*

   Using Formal and Informal Tests to Identify Instructional Pinpoints ★ Formal Tests ★ Informal Tests ★ *Using Informal Inventories to Assess Performance* ★ Creating Your Own Informal Inventories ★ Inventory Examples ★ Using Criterion-Referenced Tests

★ A Charting Shortcut for Rate Data: The Rate Finder ★ Other Charting Materials ★ *Getting Your Kids In on the Act: Teaching Children to Chart*

Static Data Procedures ★ Determining the Data Range ★ Determining the Data Mean ★ Determining the Data Median ★ Determining the Data Mode ★ *Dynamic Analysis Procedures* ★ Looking for a Line of Progress ★ Describing the Magnitude of the Line of Progress ★ Special Analysis Problems ★ The Hairpin Curve ★ Deadly Data Bounce ★ *Conclusion*

Minimum Progress Lines ★ Drawing a Minimum Progress Line ★ Using the Minimum Progress Line as a Basis for Making Program Changes ★ Seeing the Minimum Progress Line in Action ★ *Alternatives for Setting Minimum Progress Lines* ★ Using the Child's Past Performance to Define Expectations ★ Using Peer Performance to Define Progress Expectations ★ Setting a "Standard" Line of Progress ★ *Using Piecemeal Analysis: What to Do When the Minimal Progress Line Doesn't Quite Work* ★ Conclusion

Marilyn Cohen and Pamela Gross together wrote *The Developmental Resource*, a two-volume compilation of research in child development. This work, containing comprehensive reviews of the literature, as well as behavioral listings in eight important developmental areas, has been adopted as a part of the core curriculum in college programs throughout the United States and abroad.

Marilyn Cohen has conducted numerous inservice training workshops on the topics of behavior management and measurement. For six years, she served as the coordinator for Elementary and Secondary Teacher Training at the Experimental Education Unit, University of Washington. Dr. Cohen was then appointed interdisciplinary training coordinator for the University's Child Development Mental Retardation Center. She is currently the health education resource coordinator of the University's Dental/Health Education for Care of the Disabled.

Pamela Gross worked as head teacher at the University of Washington's Experimental Education Unit, before becoming assistant program coordinator for Elementary and Secondary Training there. Later, as a writer and materials specialist for the Unit, she created mixed media packages for use in inservice teacher training programs. She works currently as a freelance writer, specializing in topics concerning education and child development.

The authors express their appreciation to Greg Owen, who created the illustrations for the present work.

SECTION 1
BEHAVIOR AS A BASIS
FOR CLASSROOM DECISION MAKING

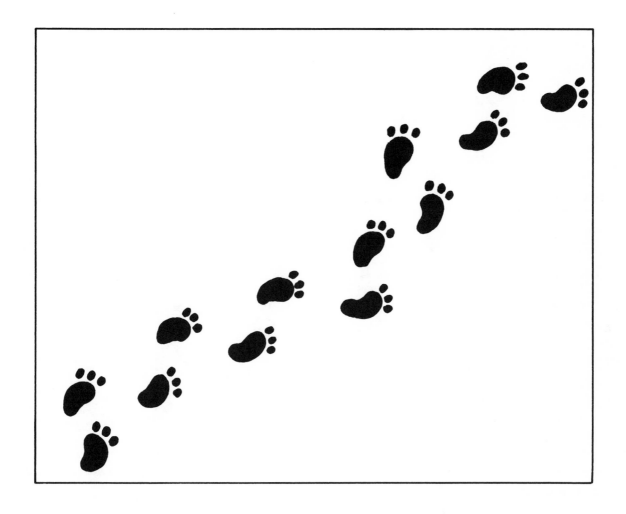

Introduction

"Every child is unique." This is the rallying cry of all educators devoted to the concept of individualized instruction. And certainly, each child is unique. Each comes into the classroom with his own history, his own family background, his own social and ethnic heritage. And, just as no child is quite like any other in appearance, in the sound of his voice, or in the way he moves, no child is the same as his neighbor in goals, interests, tastes, achievements, skill, strengths, or weaknesses.

Teachers are increasingly encouraged, and often directed, to focus on the unique educational requirements of each child in the classroom. They are asked to tailor instructional settings, materials, equipment, programs, and procedures to meet these individual needs. What is frequently forgotten, however, is that the child is not the only individual involved in the process of instruction. His teacher is an individual, too. Every teacher brings with him his own distinct view of learning and of the learner. Each holds firmly to a singular set of beliefs about how to teach and what to teach. And teachers, like their pupils, demonstrate unique combinations of strengths and weaknesses, abilities, and needs. Where do we begin an attempt to show teachers how to be responsive to the individual differences of the children in their classrooms, recognizing, at the same time, that teachers are bound to have differences, too?

This book begins with *behavior*. Behavior is something that every child demonstrates and every teacher can learn to identify. A child's behavior is a large part of what makes him unique—a sort of fingerprint in action. It is an outward expression of what may otherwise remain internal and inaccessible to us. Behavior talks; and often, behavioral messages are the clearest and most easily interpreted. Regardless of what a teacher's feelings are about *why* a child behaves the way he does, or *what* should be done about it, the teacher can pinpoint and share with others behaviors that signify a need for change. In this way, behavior serves as a neutral territory upon which the individual requirements of both child and teacher meet.

Behaviors of all kinds are instrumental in every facet of the decision-making process teaching entails. Some—raised eyebrows, frowns, or smiles—are subtle indicators we unconsciously rely upon to reflect the success or failure of our instructional attempts. Other behaviors are more blatant indicators of a child's performance characteristics and needs.

Behavior plays a dual role. By signaling the need for a change in teaching strategies, it is the starting point of our decision-making process; and, as the mirror we hold to our decisions, it is ultimately the true test of teaching success. In this latter capacity, behavior is the basis for all our measurement efforts. Because behavior is so important, the first four chapters of this book focus entirely upon behavioral issues.

Chapter 1
Defining Behavior: The Nature of the Beast

It's one thing to rely on behavior at a subconscious, informal level as a basis for decisions. A marginal amount of behavioral insight tells you whether or not you're a hit at the boss's cocktail party, and may even help you navigate tricky waters in a discussion with your mother-in-law. If you're a working mother who comes home to find the family goldfish under the sofa, you don't need a degree in criminology to determine which of your protesting offspring is the guilty party: You simply watch for behavioral signals—shuffling feet, a reddened face, downcast eyes—to help you identify the culprit.

But the decisions that fall to you daily as a part of teaching require something more than intuition. The classroom is a decision-making arena which offers no shelter to the timid. If you are lucky, the biggest problem you face in a day involves quelling a minor playground insurrection, or showing a youngster how to remember when "*i* comes before *e*." Often, however, you are not so lucky. You may confront social or learning problems on the part of one youngster, or on the part of many, all of which are more puzzling, more enduring, and more taxing. Whether the educational problems you have to solve are large or small, momentary or long term, they are important to you if teaching is important to you.

You need a basis for making teaching decisions, as well as for evaluating their effectiveness. Behavior serves as a concrete basis both for identifying learner needs and for determining how successful are the strategies you devise in meeting those needs. This chapter is intended to help you sharpen and improve your skills in the conscious use of behavior as a tool in the classroom decision-making process.

A DEFINITION OF BEHAVIOR

The definition of behavior used in this book is drawn from requirements provided by behavioral scientists:

Behavior is any readily observable action or movement that can be reliably identified.

Perhaps this definition is more restrictive than the one you would have given. You may question the significance of action or movement to the definition and wonder why we say that behavior must be observed rather than inferred. Let's examine the requirements more closely.

Behavior Is Observable: Seeing Is Believing

All of us have at least a hazy notion of what qualifies as the common, marketplace variety of behavior. After all, the words *behavior* and *behave* have been a part of our vocabulary since childhood. Think how often you have heard your parents caution: "Now watch how you behave!" If you were to ask them what they meant by "behave," chances are they would answer, "If only you weren't so *messy* (or *noisy*, or *stubborn*)."

Having grown up with it, we are inclined to feel comfortable with this concept of behavior; and, in the course of everyday affairs, we may toss out a long list of just such descriptors. We call someone assertive, aggressive, or domineering, believing that we are describing his behavior. In truth, what we are doing is something quite different than identifying behavior. We are *interpreting* behavior: that is, forming broad inferences or suppositions about someone's behavior on the basis of our own feelings concerning the actions we see them perform.

Consider, for example, the behavior of your favorite politician. You would describe him as *ethical, analytical,* and *independent.* That's why, as you read a current issue of a prominent newsmagazine, you are shocked to find this assessment of the behavior of the same man: ". . . *self-righteous, unemotional,* in a word—*passionless.*" You are astounded by the contradiction.

Yet such contradictions are to be expected as soon as we get away from specific, observable actions (for instance, your candidate's vote on each issue), and we settle for descriptors that are, in fact, interpretations of these actions. What passes for behavior in a conversation over coffee is most often seasoned with inference, interpretation, speculation, or characterization. These judgments are fine if everyone shares the same point of reference. Unfortunately, speculation, interpretation, inference, and characterization are highly personal. They are dependent as much on the viewpoint and idiosyncrasies of the individual who formulates them, as upon the actual behavior (actions) of the person they are intended to describe.

"Hyperactive," "aggressive," "antisocial," "distractible," "disruptive," and "withdrawn" are only a few of the many labels or behavioral characterizations incorporated within the textbooks and working vocabularies of psychologists and teachers. Frequently, these descriptors are a part of the professional jargon which enjoys a temporary vogue and then fades from usage as trends in the field change. These terms depend much too heavily on the interpretations attached to them. A child who is labeled belligerent and hyperactive by one teacher may seem to another merely lively and independent. To be fair to the children we work with, and to communicate clearly their needs and abilities to others who share our interest about them, we must avoid inference and speculation wherever possible.

Behavior Involves Action or Movement

One way to guarantee you are not depending on inference or other types of judgmental activities, is to describe behavior in terms of the physical movement you see. You can make the distinction between behavior and characterization by applying what we call the *Do-Test*.

The test is as simple as the distinction between *being* and *doing*. Each time you designate a behavior you want to examine, just ask yourself, "Is this something the child *is*, or something he *does*?"

If the answer is, it is something he *is*—hyperactive, messy, stubborn, etc.—it is likely that your descriptor is characterization rather than behavior. It fails the *Do-Test*. A child *is* hyperactive; he cannot *hyperact*. If, however, you have selected something your can see, something the child can *do* (i.e., leave his seat; run around the room; throw clothes on the floor; refuse to follow directions), your choice passes the *Do-Test*. You have targeted a behavior.

USING A BEHAVIORAL RULER

The process of identifying and describing a behavior you intend to work with and change is called *pinpointing*. While pinpointing is accomplished in a variety of ways, its goal is always the same: to identify a behavior which describes as exactly as possible what the child must do in order to succeed in the classroom, and in the world outside it.

Behaviors come in all shapes and sizes. Some are broad, all-encompassing categories of activity. Others are extremely precise, minute, fussy creatures. To say that you want a child to read well is to identify a behavior of the first kind; to say that you want the same child to increase his accuracy in reading orally words with a c-v-c (consonant-vowel-consonant) pattern, is to pinpoint a behavior of the second sort. Deciding what degree of precision is right for you will be much easier if you think of behaviors as falling along what we call a *behavioral ruler*.

A linear ruler, such as a yardstick, is divided according to a variety of increments, which progress downward in size from yard, to foot, to inch, half-inch, quarter-inch, and eighth. Each increment allows us to refine our description of an object's linear dimensions. The smaller the unit, the more precise the linear description. Sometimes a gross measure is sufficient. Figuring out the size of a room you intend to paint, for instance, allows you some leeway. A few inches more or less won't make that much difference. Determining dimensions to remodel your house and add a new room onto an old one, however, requires much more exacting measurements. You'll need information calibrated down to the smallest fraction of an inch.

Behavior, too, can be viewed in terms of a scale whose units range from large to progressively more refined. Depending upon the requirements of the decision-making situation, you may choose to deal with a very gross behavioral unit, called a *chunk* in this chapter; or you may select a smaller unit, such as the *movement cycle*. The size and specificity of the behavioral unit depends entirely upon the nature of the learning question you are trying to answer. That's why we talk about determining a *functional unit of behavior*.

Behavior Chunks

Behavior chunks are the largest, least refined unit of behavior you can use to describe observable action. They're like a huge, delicious piece of cake that you take at a swallow. The term is

Getting Ready to Apply the Behavioral Ruler

one we have coined and given the following definition:

Chunks designate gross, broad-reaching actions that may incorporate a number of smaller, more discrete behaviors.

Paying attention, showing cooperation, throwing tantrums, and *following directions* are four behavior chunks that teachers, and parents especially, frequently *show concern* (another chunk) about. It's not difficult to guess *roughly* what these chunks entail, but it's likely that the individuals who establish the increase or elimination of these behaviors as a priority will have slightly different definitions in mind. This is the tricky part about behavior chunks.

Because a chunk may incorporate a vast number of smaller behaviors, in order to avoid confusion it is necessary to define each chunk clearly, keeping in mind the specific circumstances in which it will be applied. The definition you provide must be clear to you and to whoever else will be involved in monitoring and working with the behavior in question. You are free to make that definition as simple or as complex as the situation requires.

Defining chunks with simple statements. The easiest description is, of course, merely a short statement defining the behavior in the simplest terms possible. The statement should be specific and clear enough that any other person would be able to identify the behaviors easily as soon as he saw it. As you formulate such definitions, keep in mind, also, whether or not the definition allows you to explain the behavior easily and clearly to the child himself.

Take a behavior such as *showing cooperation.* This is a chunk of considerable interest to parent, teacher, and employer alike. It is also a chunk of considerable behavioral magnitude. To explain what you mean by *showing cooperation*, you might simply extract one of the definitions found in any good dictionary and modify it to suit the setting in which cooperation is desired. "Timothy *shows cooperation* by working together with others to accomplish specified tasks," is one example.

To determine whether or not your dictionary definition is sufficiently precise, apply this test: Ask yourself whether two or more people viewing this behavior, and going only by the definition you've given, would agree that the behavior *shows cooperation* had, indeed, occurred.

Defining chunks by listing component behaviors. Even a definition drawn from a dictionary may be subject to some ambiguity. You can achieve an even clearer definition if you are willing to list the smaller, more discrete behaviors that comprise the chunk you have in mind. A child whose problem behavior is described as *throwing tantrums* for instance, may indulge in a variety of behaviors which include:

— pouting
— crying
— whining
— swearing
— kicking
— hitting
— biting
— running away
— tearing up materials
— throwing himself on the floor
— throwing desks, peers, or even teachers to the floor

Anyone watching a youngster engage in even one of these behaviors in isolation would probably agree that the child was, indeed, having a tantrum. But certain questions remain about the total scope of this problem. Eventually, you will establish a plan to deal with the behavior you have pinpointed. Before you reach the intervention planning stage, however, you need to make sure that the behavior's definition is clear enough to allow you to intervene consistently each and every time an instance of it occurs. Consider these questions:

— Does the child have to engage in a combination of the listed behaviors at once in order to be considered as having a tantrum, or is something like *whining* enough in itself to qualify as a tantrum?
— Are each of the smaller component behaviors weighted equally? That is, does each have the same nuisance value; or would you consider some more annoying, or more potentially dangerous, than others? Would you, as a teacher or parent, be willing to ignore a little pouting or whining—maybe even a little outburst of crying—but definitely draw the line when physical violence erupts?

Here's how one teacher handled this behavior chunk:

Sheila Dodd found these questions to be relevant in her work with Randall, a thirteen-year-old pupil referred to her intermediate special education classroom for a perplexing collection of behavior problems which included throwing papers, books, and chairs, kicking, hitting, spitting, crying, and whining. Because they appeared to be related, Sheila grouped these behaviors together under a single, easy-to-count designation: *throwing tantrums.* Whenever any one or a combination of the behaviors constituting "tantrums" occurred, Ms. Dodd carried out the consequence she had decided upon in agreement with the school's psychologist and Randall's parents—removing Randall from the classroom setting until he calmed down sufficiently enough to return to work. Although this "time-out" procedure was a drastic intervention, the severity of Randall's tantrums seemed to warrant it.

As the program took effect, Randall's tantrum behavior did, indeed, begin to change. The content of the tantrum episodes slowly altered: Throwing, kicking, and hitting began to fade, while—simultaneously—Randall seemed to compensate by crying and whining more. With all the other things going on in her classroom, however, Sheila didn't really notice the subtle shift in behavior that was beginning to occur. And her data, based solely on the chunk *throwing tantrums*, did not reflect the gradual qualitative change, either. The data, in fact, showed tantrums to be continuing at a fairly constant level, when, in reality, the nature of the tantrum episodes was drastically different, and certain parts of Randall's tantrum behavior had decreased significantly.

At the same time, Ms. Dodd was missing an important opportunity to find a consequence for whining and crying—relatively mild behaviors compared to the others—that was less severe than the time-out procedure.

Remember: If you are going to intervene in an effective and timely manner to change behavior, you need to be certain enough of the definition to allow you to act immediately and consistently. By lumping several behaviors together, as you do when you list the components of a behavior chunk, you inevitably lose some of the accuracy and sensitivity critical if you decide to take behavior data. You need always to ask yourself whether or not the final data will represent the picture of change you suspect you are seeing, and whether or not your records will reflect this behavioral change in a way that makes sense to others.

Make sure, before you do anything else with a chunk, however carefully listed or defined, that the behavior you have chosen is reduced to the size you want it to be. There may be additional distinctions and discriminations among behaviors important enough for you to want to break your behavior down still further. If the questions raised so far generate more questions on your part, you may want to consider still another kind of behavioral definition.

Developing a functional definition of behavior. A third way to clarify what is meant by a chunk is to give it a "functional" definition.

A functional definition of behavior describes how a behavior functions in terms of the environment. The definition includes specification of at least one of the following criteria:

1. The behavior's physical or spatial limitations;
2. The behavior's temporal limitations, or how much time is allowed;
3. The changes the behavior will make on the environment.

Consider a chunk frequently identified by teachers as a behavior they would like to see enhanced in the classroom, *attending to task.*

Imagine that Fatty Fenimore is a particularly troublesome kid in your reading class whose attending behavior you desperately want to shape up. You break out what you call "attending" in this case, using all three specifications needed for a functional definition, and *voila...*

Behavior
   Attending during reading period
Functional definition
1. Physical limitations
   a. Fatty sits at his own desk
   b. Faces forward
   c. Keeps his hands on his desk; does not play with pencils, erasers, etc.
   d. Focuses eyes on the page in front of him
   e. Eyes move left to right across the page
2. Temporal limitations
   a. Continues to read "silently" for entire twenty-minute study period
3. Changes in environment
   a. Turns pages as he finishes them
   b. Writes answers to comprehension questions on worksheet

You're pretty proud of the job you've done, defining this difficult and chunky behavior on the part of a "difficult and chunky" boy. But Mr. Andrew Adidas, Fatty's gym teacher, raises an objection. He's concerned with Fatty's attention, too. But he certainly doesn't want the kid sitting in one place for twenty minutes. And he doesn't really care whether or not Fatty is all that quiet. As for his hands, using them to play is the most important part of a basketball game, except for the footwork, maybe. When Fatty is attending to what's going on in the gym class, he'd better be a little more lively about it than what you've specified!

In this case, each teacher involved will have to come up with his own functional definition of *attending to task.* Spelled out in this way, the behavior being attended to is absolutely clear to all parties responsible for behavioral change—Fatty's teachers and Fatty himself.

Granted, functional definitions are clear and precise, but their very precision may pose certain problems. Frequently, the specifics of a definition will differ markedly, depending on the setting. Further, certain criteria may fluctuate over time. For instance, as Fatty gets better at attending, you may want to increase the time requirements to 30 or even 40 minutes; at the same time, you may become less rigorous about some of the accompanying components, allowing him to talk quietly to a neighbor, as long as the specified number of comprehension questions is answered correctly by the end of the reading period. Such changes will have to be accounted for by an alteration of the behavior's definition each time they occur.

Again, it's important to realize that what you have defined in functional terms as *attending to task* is specific to the task at hand. What Fatty's mother means by attending, or what another teacher has in mind, may be entirely different from your conception. In these circumstances, data describing Fatty's attending behavior across several different settings may be difficult to compare. Although a behavior called *attending* is measured in each case, its functional requirements are not uniform from one setting to the next.

Movement Cycles

Sometimes, chunks simply do not give you a precise enough way to describe the behavior you want to measure and change. If this is the case, you may want to identify behavior in terms of movement cycles. These are the smaller, more refined units on the behavioral ruler. Before giving you a definition of the term *movement cycle*, we'll let you look at some yourself and compare them to the behavioral chunks you have just been dealing with.

As you examine the three following lists, try to determine for yourself some of the differences between chunks and movement cycles, and the implications these differences might hold for you.

FIGURE 1:1. *Behavior Reduction Plan.*

Mega-Chunks	*Chunks*	*Movement Cycles*
to understand	to differentiate	says numbers
to know	to relate	writes numerals
to appreciate	to identify	writes letters
to grasp significance	to compare	marks answers
to enjoy	to solve	points to item
to demonstrate	to construct	writes words
to comprehend	to contrast	says answers

As you compare the lists, it becomes obvious that the behaviors in them differ chiefly in terms of the clarity and precision with which they identify the movement or action to be observed. List 1 contains the behavioral "biggies," open to the broadest of interpretations, and consequently, misinterpretations. You'll recognize these behaviors as a collection commonly found housed in curriculum guidelines. Let's take just one of these behaviors—*to grasp the significance*—and reduce its dimensions slightly by subjecting it to a kind of behavioral diet and exercise plan we like to call the Behavior Reduction Plan.

How would a teacher, asked to ascertain that his pupils *grasped the significance* of concepts and operations introduced in the course of study, do just that? He might start by asking students to perform any one, or even a combination, of the behaviors found in List 2: for example, having them *solve* problems; *differentiate* concepts; *relate* instances; *identify* opinions; *compare* quantities; *construct* theories; or *contrast* ideas. Suppose, out of this list, he settles on *solves problems* as a behavior that will let him see how completely a youngster grasps the significance of program content.

Identifying such a behavior gives the teacher a somewhat firmer grasp on exactly what it is he expects to see when he attempts to determine whether or not students are performing adequately. And yet, the precise nature of the hoped-for behavior is still open to considerable question and conjecture. What, specifically, is it that the child does to show he *solves*? The actions that constitute even this single behavior may differ widely, depending upon the requirements of the individual teacher or individual content or concept area. That's because each of the chunks listed in List 2 actually designates a *class* of behavior, rather than pinpointing a specific action the teacher can observe, count, and record.

The behaviors in List 3, called *movement cycles*, represent the final stage in the Behavior Reduction Plan. These behaviors come without any excess dimples or curves—they are pure movement, and there's no guesswork necessary about what they entail. If a teacher says he'll look at *writes numerals* behavior in order to tell how well a child is able to *solve thought problems* involving fractions, there should be no difficulty on anyone's part envisioning exactly what it is the teacher wants the youngster to do. Just as any of the chunks in List 2 might be used to describe the behavior to *grasp the significance* from List 1, so any single movement cycle or a combination of them from List 3 can be pinpointed to describe the way in which a child shows his ability to perform a behavior chunk from List 2. This is what we mean by refining and reducing behavioral pinpoints: trimming off all the vague and hazy terminology to find the precise behavioral description underneath.

Movement cycle is a term originally devised by professionals who created a set of measurement procedures collectively called *Precision Teaching* (Kunzelmann et al. 1970, Lindsley 1971). While the term may seem to you at first somewhat strange, or even whimsical, it is an entirely functional one.

> *A movement cycle is a discrete behavioral unit. It contains a single behavior (movement) which has a definite starting and stopping point (cycle). At the end of the cycle, the movement is able to begin over again, so we may say that the movement cycle is repeatable.*

A single, discrete movement. As we have already demonstrated, when we talk in terms of behavior chunks, every chunk identified could actually include a number of smaller behaviors—each of which might be very different from the others. *Having a tantrum*, for instance, might include *crying, throwing things* on the floor, *hitting, kicking, pulling hair*—any one of these behaviors, or perhaps all of them combined.

Each person you talk to would probably give you a different description, a slightly different list of behaviors, if you asked him to describe *throwing a tantrum*. We say that such a chunk may include a variety of behaviors with different "topographies." That is, the configurations, or outlines, or shapes of each of the behaviors that might be included in the chunk, are different.

But a movement cycle includes one, and only one, specific kind of movement or action. It has a unified topography. If you say, "I am going to look at how often Ralph *hits others*," you are talking about a discrete movement: *hits*. There is no way that *hits* will include cries, throws, bites, or kicks. Anyone watching Ralph *hitting others* would immediately identify the behavior, in almost identically the same terms. If you asked someone to count each time he observed the behavior, that individual would be more than likely to know exactly what he was looking for. This kind of clarity simply isn't possible with a behavioral description such as *throwing a tantrum*, where any one of a number of discrete behaviors, or a combination of these smaller behavioral units, may be included in the definition.

A definite starting and stopping point. The second element in the definition of the movement cycle—the *cycle* part—is also extremely important to the clarity with which movement cycles allow you to identify, measure, and communicate about behavior. The cycle will be especially critical when you are ready to count behaviors. By *cycle*, we mean that the behavior has an easily

Movement Cycle in Action

identifiable starting and stopping point.

Look again at the example of *hits others*. This movement cycle has a very marked beginning and ending point. The movement starts when the child raises his hand to begin hitting. It ends when physical contact is made and the hand is subsequently withdrawn. You are back at the starting point, behaviorally; the child is now able to engage in the behavior again, or to take up a completely different, competing behavior with his hand. Each strike counts as a discrete movement.

Since the idea of *cycle* may be a new one, let's try it on a few more behaviors. Consider the following commonly pinpointed movement cycles:

FIGURE 1:2.

Movement	*Cycle*
Takes bites	The movement begins when the child opens his mouth; ends when the teeth close on bite of food.
Gets out of seat	Begins when child's body leaves bottom of chair; ends when child returns to seat.
Writes letter	Begins when pencil or pen touches paper to begin writing; ends with finishing stroke of letter (e. g., writing the letter *a* counts as one movement cycle).
Writes numeral	Same as above, with a numeral as the product (e.g., *4).*
Shares toy	Begins as child hands toy to other individual; ends as toy leaves child's hand, freeing him to begin new action.

You'll notice that, in order to define the movement cycle, you must either explicitly or implicitly take into account the movement's interaction with an object.

Action	*Object*
Takes.	.bites of food
Gets out (of)	.seat
Writes	.letter
Writes	.numeral
Shares	.toy

The pinpointed action ends—the cycle is completed and ready to begin again—when the movement completes contact or interaction with the object. We have stressed already that a behavior is something an individual *does*. Movement cycles take this requirement one step further: They describe not only what an individual does, but also what he does it *to*.

As you gain skill in identifying, or pinpointing, movement cycles, you may find yourself using an abbreviated form to describe the cycle. Sometimes you may identify a movement cycle in which either the "object" or "movement" part of the pinpoint is clearly implicit, but not necessarily explicitly stated. *Smiles, out-of-seats, pouts,* and *talk outs* all readily come to mind. These pinpoints can all be stated formally in such a way that they include both movement cycle components just delineated—for instance, *makes smiles* or *makes pouts.* This is not, however, the way in which we speak of these activities colloquially; and, phrased this way, they sound cumbersome and a little silly. If the meaning is clear, you need not always spell out the movement cycle in full. What is important, is that both movement and cycle are clear to you and to anyone else who will be monitoring or working with the child's behavior.

Movement cycles are repeatable. Once the cycle is complete, the movement is able to resume again. It is important that movement cycles be repeatable, since behavior cannot change if it occurs only once. The more frequently a behavior is likely to occur, of course, the more chances you will have to measure and interact with it, affecting change. For this reason, it is advisable to pinpoint movement cycles that the student is able to produce fairly frequently.

As the teacher of a fourth-grade science class, for instance, you might find that a major goal identified by your district administrators is that "students will demonstrate increased curiosity about the relationships of natural elements and forces in the surrounding environment."

You have already clipped some of the thorns from that directive by reducing it to its behavioral statement: *demonstrates increased curiosity.* Just the same, you are aware that this particular chunk comprises many smaller behaviors that vary greatly in relevance and applicability from one setting to another, as well as from one student to another.

You might, for example, require that students *complete outside reports, work supplemental problems,* or *read additional books and articles.* Each of these is a clearly identifiable and observable behavior that might be taken as a demonstration of curiosity. Yet, each of these is a large chunk of behavior, requiring a vast collection of smaller skills. It might take a student weeks to complete any one of the tasks; and, if that student is able to prepare only one extra report a quarter, you as his teacher will have a difficult time determining how much his curiosity level has changed over the academic period.

By selecting a smaller behavioral unit, a movement cycle such as *asks questions,* you give the student a chance to repeat, or practice, the behavior many more times. You increase his opportunity to develop curiosity, because you have pinpointed a behavior that can be repeated and encouraged again and again. At the most, a fourth-grader would generally read one extra book a week; but he can ask numerous questions in the same amount of time.

A movement cycle's repeatability is important to keep in mind as you pinpoint given behaviors. The more frequently a child can repeat a response, the more opportunities he has to practice that behavior, and the more opportunities you have to interact and change it.

CONCLUSION

Life in the classroom throws you one demanding situation after another and asks you to do your best with each. In order to solve learning problems and answer instructional questions effectively, you must be able to identify and describe the problem with precision and clarity. By specifying the behaviors with which you want to work precisely, you take an important first step.

As you progress through this book, you will continue to develop and refine your skills in isolating behaviors which are crucial to the child's progress. At the same time, you will add to these skills other measurement techniques, which will allow you to assess the child's success as a learner and to evaluate your success as a teacher.

Chapter 2
General Pinpointing Strategies

Putting your knowledge of behavior to use in the classroom means that you must be able to identify those behaviors which are critical and functional: behaviors that are critical to each child's successful classroom performance, and that bear a direct and functional relationship to overall educational objectives. It would be nice if you could pull these pinpoints out of the air, or simply draw them out of a hat. Unfortunately, pinpointing is seldom a matter of divine inspiration. Each youngster you work with is likely to present an entire fistful of behaviors which demand your attention, and no one has to tell you that you would be hard-pressed to work with every one of them at once. How do you choose the specific behavioral targets you will use as a basis for decision making?

GETTING STARTED: INITIAL GUIDELINES

Shopping for the right pinpoint is an endeavor which falls somewhere between trying to find the perfect outfit and buying a week's food for a family of five. Approach it as you would any foray into Bonwit Teller, or your local supermarket: with a healthy mix of daring and caution. As every experienced shopper knows, certain rules make any shopping trip a more profitable experience.

Make a Shopping List

Shopping's always faster and more economical if you can decide ahead of time what you need. Generally, writing up a shopping list helps. You plan purchases based upon a survey of existing conditions and provisions in your pantry. Pinpointing can be conducted in much the same

In the Market for Behavioral Pinpoints

way, working from a list or framework drawn up before initiating programs. Start by defining the forces that influence what you do in the classroom.

What are the outside forces affecting your decision making? Administrative directives; district guidelines; program graduation or referral requirements; requirements imposed by an Individualized Education Plan (IEP); and informal input from colleagues or parents—all of these influence the programming decisions you make and the types of pinpoints you decide to look at.

What is the nature of the curriculum selected? What kinds of books and materials will you be using? Are behavioral objectives already defined in regard to these materials; or, will you be determining them for yourself? Do you plan on developing individualized programs, or on using a uniform program and set of instructional materials for all the youngsters? Maybe you'll mix and match, implementing individual plans when they are necessary, based upon the child's performance.

What are the child's needs, and how are these identified? Do you plan to implement formal evaluation procedures? Will you use informal tests and inventories? How much do these identified needs direct your choices in the selection of instructional materials, techniques, and program changes?

What is your own personal teaching philosophy? Teachers, just like children, come into the classroom with their own sets of needs and expectations, interests, and experiences. Your personal teaching philosophy and style will shape significantly the kinds of behavioral pinpoints you emphasize as you design and implement instructional programs.

Whatever you're shopping for, whether groceries or pinpoints, never wait until the last minute. Going into the market hungry, means going in desperate. You'll be at the mercy of every sales person you trip over. It's impossible, of course, completely to avoid emergencies in the classroom. Unexpected situations crop up in any human endeavor, often requiring the brave implementation of not-too-well-thought-out plans. But on the whole, planning ahead is the key to success. Your pinpoints will stand a much better chance of satisfying instructional needs if you can identify and describe those needs well in advance.

Decide on a Shopping Strategy

Wandering aimlessly up and down the aisles, or in and out of departments, will cost you time and money. Just as important as knowing what you want before you enter the store, is looking for signs that can direct you to the right areas and items once you get inside. A shopping strategy is also crucial if you want to arrive at behavioral pinpoints which have long-term applicability.

Set up a structure ahead of time that will give direction to your day-to-day plan making. Whether you intend to follow district guidelines to the letter, or whether you plan to supplement administrative directives with a variety of materials and strategies you yourself have chosen, outline some kind of basic framework that will suggest content areas and the behaviors which best indicate content mastery.

A variety of materials can act as frameworks for generating important behavioral pinpoints: standardized or informal testing tools; behavioral checklists and sequences; textbooks; teachers' manuals; the lists of goals and objectives in a curriculum guide; authoritative books and articles; the resource people available within the school itself or the community at large. Decide which of these sources of information and direction you plan to use. Once you have identified and analyzed these frameworks, you can begin to draw pinpoints from them.

Shop Comparatively and Competitively

If you know the comparative cost of items and the restrictions of your own budget, you're more likely to get the best value for your money. The same kind of comparative, competitive shopping is a must when you're in the market for pinpoints. Decide on behavioral pinpoints carefully. Make sure they describe the behavior you want to work with accurately and sensitively. If the pinpoint you choose now seems inadequate later, don't discard it without considering how you might change it so it will work, or what alternative pinpoint might better suit your needs. Equally important, know the *cost* of the behaviors you pinpoint, in terms of program time and effort required of both yourself and the child you teach. Sometimes a pinpoint appears sound initially, but in the end, the results don't seem quite worth the energy and frustration teacher and pupil expend. Know when to compromise and readjust your goals.

Avoid Impulse Purchases

When everything on the shelves looks so tempting, sometimes it's hard to stick to your list. But remember how it feels to go home and face the donuts you don't need, or the closet full of clothes you never wear. It's better to end up with one good pinpoint than with an entire cartload of behaviors that are not quite right. Try to make sure that each pinpoint you select does the very best possible job of mediating between outside forces and requirements, the demands of the curriculum, and the special learning needs and characteristics of the youngster. No program or instructional technique will give you the results you're looking for if you are careless in defining how you want a youngster to show that he is learning. *The pinpoint is the whole point*, so pick the best one you can.

Establish Pinpointing Priorities

If you had the time, you might want to identify several behaviors for each child—some academic and some social—to which you would devote special attention (at the least, giving them a "spot check" every week or so to make sure things were heading in the direction you wanted). Unfortunately, you probably don't have that kind of time. You make a note to yourself, that one child seems to have difficulty making friends, that another seldom contributes to class discussions, and that a third is so far ahead of the rest in most areas, she's probably getting bored. How do you decide which of these little and not-so-little problems merit immediate action? The four issues discussed below should help you establish much needed pinpointing priorities.

Will this behavior result in the child's removal from his present educational situation? Behaviors which are likely to lead to the child's immediate or eventual placement in a more restrictive setting—such as transfer from a regular to a special education classroom—are generally the first targets for measurements and change. These behaviors are usually called "social" or "antisocial": hitting peers, stealing, and destroying property, are some examples. But often, other difficulties prohibit participation in regular classroom activities; perhaps the child exhibits math or reading deficits which will eventually lead to special class placement.

Many times teachers find that, although a child's problems seem overwhelming, the major reason for removal can be traced to only a few areas of concern. A single behavior, such as *talking out*, may so overshadow the child's other redeeming features, that the teacher cannot contemplate keeping him in the classroom. Ask yourself what changes in behavior would make it possible for the child to continue in your classroom. Try listing positive and negative points, and consider what behavior change would make the most immediate and dramatic difference to you. Sometimes teachers discover after working with a problem, such as *out-of-seat*, that they are better able to handle the child within their own classrooms and change their minds about the necessity for referral.

Precise pinpointing will help you obtain more immediate and exact help from other interested professionals who can aid you in meeting the child's needs within your classroom, or in making a better placement for him based upon the problems you have clearly identified.

Is the behavior crucial to the development of "survival" skills? Depending upon what the child brings to the classroom, a second category of behaviors, called "survival skills," may have first priority. Suppose you are working with an 18-year-old moderately handicapped youngster who has not yet acquired basic reading and arithmetic skills. You will probably want to forego the traditional reading or arithmetic program in favor of one which stresses reading survival words such as "Danger," "Men," "Women," "Exit," along with focusing upon skills such as making change and telling time. For the growing number of "latchkey" children who return from school to an empty house, skills that will enable them to take care of themselves in the hours between school and their parents' arrival home take on the same survival value.

Older youngsters, preparing for entry into the job market, may desperately need instruction in filling out employment applications, handling job interviews, and developing other skills which will help them keep a job once they find one.

Is this behavior one which allows the child's continuing progress or promotion? Next in line are those behaviors crucial for moving from one level of performance to the next: from one reader to the next; from one grade to the next; from a resource room to a regular classroom; or from school to the work environment.

Is the behavior part of an enrichment activity? These behaviors are usually the ones left for last. Although they are a vital part of growth in a child's responsiveness to the world around him, they are not often given priority in the curriculum. This is a generalization which does not, of course, apply accurately to all programs, all teachers, or all children. Even if "enrichment" is last on your list of priorities, it is important that you at least consider it, whether or not you have the opportunity to measure behaviors essential to performance in "the arts" or enjoyment of "leisure time."

EXPLORING PINPOINTING SOURCES AND STRATEGIES

By selecting pinpoints that relate to a preestablished framework, whether formal or informal, you are more likely to guarantee that the behaviors you focus on will ultimately result in a comprehensive program of instruction. The following represent major forces which direct program planning and the identification of behavioral targets. Each acts as a kind of framework, suggesting behaviors that are necessary for functioning successfully in the classroom. And each entails a slightly different strategy for pinpoint identification.

Brainstorming: An Informal Framework

The least structured way to arrive at a pinpoint is to watch the child himself as he engages in a variety of both structured and unstructured classroom activities. Then brainstorm, identifying "off the top of your head" what behaviors seem indicative of social or academic deficits, as well as behaviors which seem significant to success in the programs you have established. The frame of reference to which you compare the child's behavior is your own general feeling about what is appropriate, meaningful, or necessary.

This informal procedure is most frequently used to pinpoint behaviors falling into a category labeled as "social." Since detailed standards for social performance are not generally a part of curriculum guidelines, it is usually left to the teacher or the school psychologist to pinpoint those behaviors which most adequately describe a social problem. Every teacher knows what behaviors he can live with, and what behaviors absolutely demand intervention. Many teachers are able to tolerate a certain amount of classroom chatter, students leaving their seats, and even a paper airplane or two; but it is unlikely that a teacher will let hitting, yelling, or throwing food happen more than a few times without planning some immediate intervention.

When a youngster in your classroom repeatedly displays behavior which interferes with learning and classroom communication, or which is potentially dangerous, it is quite obvious to you that you need to do something about that behavior. Pinpoint a movement cycle which comes closest to the mark in describing the problem behavior. If records forwarded to you indicate that an incoming pupil has had problems in his previous placement, try listing these problems, giving them the best behavioral description you can, and check the list against what you see during the first weeks of the child's adjustment.

While identifying pinpoints in an informal, unstructured way works adequately enough for those behaviors which "hit you in the face," so to speak, you may find a more structured approach necessary when it comes to identifying academic pinpoints.

Expert Opinion

Another fairly unstructured source for pinpoints lies in the advice of outside authorities. Questions concerning how to deal with an unusual or perplexing problem may lead you to seek outside opinion. You may contact a fellow teacher, the principal, school nurse, or school psychologist; or you may request the help of a resource person from the local school district, college, or university. Even if these individuals don't actually assist you in pinpointing behaviors as a starting point in problem solving, you might be able to draw pinpoints on your own from the suggestions and material they give you.

Books and educational journals pertinent to a particular skill area or learning deficit serve as an excellent additional resource from which you can extract behavioral pinpoints which apply to issues you have identified in your own classroom. Research reviews are an especially good source of behaviors which experts have gone to considerable lengths to identify and verify as important to specific educational populations or skill areas.

As you read or talk to authorities, consider these questions:

— Does the authority suggest precise behaviors to look for? Does he specify the context in which you would be most likely to see these behaviors occur?
— Are the behaviors stated with sufficient clarity that you can use them in your work; or, will you need to implement a Behavior Reduction Plan?
— Do the authors' suggestions seem to rely on one movement cycle more than another— such as *say* or *write*? Or do they suggest tasks calling for a variety of behaviors? Which behaviors seem the most critical?
— What agreement about behaviors do you find across sources? Often, books written to

describe a single instructional area or educational population (e.g., books written about language development or about autistic children) show considerable overlap in the behaviors they mention as relevant to the topic. These behaviors would certainly be ones you would want to emphasize in your own work.

Requirements Dictated by Administrative Guidelines

Frequently, administrative requirements, whether formal or by word of mouth, define overall goals toward which child performance should be directed. These goals and guidelines may or may not be behavioral in nature. Quite often, they are rich in behavioral "biggies"—such as *to understand, to comprehend, to develop,* and *to appreciate.* In order to reduce these chunks to a more concrete level, you may ask yourself:

— How much guidance am I being given in selecting pinpoints, and what restrictions must I keep in mind?
— What are the given environment and materials; how free am I to choose pinpoints in relation to these restrictions?
— How do whatever restrictions under which I must operate relate to the children involved: How can I pick behaviors which are sensitive to their needs and, at the same time, meet administrative guidelines?
— What behaviors will I show as proof that the child has performed according to the guidelines or requirements I am given? What are the most easily identifiable and measurable behaviors that will show me—and others—that the child is demonstrating the called-for understanding, comprehension, appreciation, etc.?

Instructional Materials, Textbook and Workbook Units

To determine what behaviors best indicate successful academic performance, the quickest and most direct route generally lies in consulting the instructional materials in which you actually expect performance. As you scan these materials, it should be obvious that the activity we commonly call "learning" involves both a behavior and a content component. Although it is easy to confuse the two, in order to pinpoint successfully, you need to be able to separate one from the other. *Content* is what you expect the child to learn; *behavior* is how the child demonstrates that learning has occurred.

Content may be defined in terms of broad skill areas such as (in reading) comprehension, word attack, and vocabulary development; or, it might be related to specific units to be presented, such as those which might be defined in a science program, covering animals, plants, magnets, and simple machines. Although the content may show considerable variability, the behavior selected to signify performance of these tasks might be a single movement cycle, such as *say words.*

You can reduce and simplify the job of determining what behaviors signify learning within instructional programs significantly by *first* identifying the program's content and any sequence that content might fall into. *Then,* go back through the sequence you have constructed and decide what behaviors are called for. These behaviors may be stated explicitly in the text or manual, or they may be ones you have to identify. Sometimes you will find that the same behavior can be used throughout an entire sequence; at other times, the behavior will change as the material changes. From now on, whenever we talk in this book about pinpointing in relation to instructional areas, sequences, or materials, we will assume that you understand that behaviors are identified in relation to, but separate from, specified content.

Depending upon the way the instructional material you are using is organized, there are a number of strategies which will help you begin pinpoint selection:

— You could start by referring to the teacher's manual, where you may find a ready-made list of objectives or goals. These can be turned into pinpoints for measuring pupil performance by extracting the behavioral component from the stated objective.

— Consult any commercial test materials which accompany the textbook. These should give you some indication of the type of performance expected.

If these strategies fail to provide useful pinpoints, you will have to resort to analysis of the text itself.

— Start by listing the skill sequence or content areas. Next, list all the major behaviors or responses that seem to be related to each content area or step in the skill sequence. Then make a tally to determine what type of response(s) seem to be most crucial or consistently required throughout. The behaviors which appear most weighted as a result of your tally will probably be those you want to choose as a measure of task accomplishment.

— If you fail to find any, or only a small number of behaviors which seem of overall importance, decide for yourself what responses represent the best match between learning requirements and the individual pupil's performance capabilities.

Sometimes a text or workbook is so poorly organized that it defeats the most earnest attempts to isolate a clearly defined instructional sequence, as illustrated by the following vignette:

Consider the problems experienced by Ernesta Little, who had just begun her first year of teaching at Hillbrook Elementary. She was expected to use the math materials that the district had developed over several years. Scanning the workbook her new first graders would face, she identified a list of skills as seen in Figure 2:1.

FIGURE 2:1. *Ernesta's Skill Listing for Math Workbook.*

Determining size-shape relationships
Finding 1:1 correspondence
Arranging sets in order
Counting pictured objects (0-10)
Counting dot configurations (0-10)
Matching numeral with set
Copying numerals from sample
Pairing numerals with words (0-10)

Ernesta was concerned when she observed that sometimes four to five different tasks appeared on a single page. Not only did content change rapidly and repeatedly, but the behaviors called for included *drawing lines, circling pictures, writing numerals,* and *cutting* and *pasting.* Ms. Little feared that the youngsters might experience frustration because they were asked to cope with such frequent task changes, and that they also never received adequate repetition on any one task to allow acquisition of the intended skill. The fact that the behaviors called for changed frequently, compounded the confusion.

She decided that the easiest way to deal with this problem was simply to take the workbook apart, rearranging and reorganizing the tasks into clusters of smaller task content. Tasks were further regrouped according to the number of items in a set, resulting in a logical progression from zero to 10.

Now she had created worksheets which contained uniform tasks, but which might demand a number of different movement cycles. In order to make the tasks still more uniform, she decided to use a consistent movement cycle within each of the task groupings. Thus, she could present task types in an order she felt corresponded with the development of basic math skills (e.g., *size-shape relationships* and *1:1 correspondence* preceding *counting pictured objects*).

These new and better-sequenced worksheets could then be copied in the easiest and least expensive way for distribution to each child.

Formal and Informal Evaluative Tools

Results from both formal and informal assessment procedures often suggest behavioral pinpoints that require further instructional attention. Chapter 3 details the use of such results for pinpointing purposes. Pinpoints can also be drawn directly from the assessment instruments themselves. If, rather than implementing an assessment tool, you plan merely to scan one or more of such test instruments to extract possible pinpoints from them, you may find the following hints helpful:

— Survey test items and extract behaviors called for. Sometimes behaviors which are significant from the test designer's point of view for "getting at" specified skills or concepts, will need changing in order to be useful for examining skill development within your own classroom environment. For example, the WRAT—*Wide Range Achievement Test*—contains a spelling subtest which requires written responses. A teacher may consider the test to be a valid measure of spelling skills and strategies, but find that a pupil in his special education classroom is unable to write letter responses with enough ease to allow the child to complete the test. What alternative movement cycles might be employed to examine this important skill more closely?
— Judge whether the behaviors called for are chunks or movement cycles. If test items call for such large chunks of behavior that you suspect you will have trouble sorting performance pluses or minuses of a specific kind from the total chunk, you may need to apply the Behavior Reduction Plan.
— As you analyze the test or inventory, determine whether you find any behaviors that keep reappearing. Note, as you survey the test material, what behaviors—if any—seem to be required again and again. In some inventories, a single behavior is used to examine increasingly complex, and even diverse, skills. For example, the *Classroom Reading Inventory* (CRI) (Silvaroli 1973) examines both word recognition and reading comprehension skills over a range of grade levels, by requiring a single response type, *say words*.

Skill Sequences and Checklists

Skill sequences and checklists come in a variety of forms, depending upon the rationale behind their construction (e.g., developmental versus task analysis approaches).

It is our belief that no child develops precisely according to the timetable or pattern laid out by even the best researched developmental sequence or scale. The instrument itself, however, offers an invaluable overview of what can be expected as a general sequence of skill acquisition. That is, it allows us to see the sequence of skills in terms of content changes, and to correlate these changes with behaviors that would best describe the child's performance at each step. Such an overview helps to remind us of overall program goals. This is a function especially appreciated by teachers who must often break tasks into such minute pieces that they frequently forget what the whole behavioral jigsaw looks like when it's put together.

Individualized instruction is usually most successful when behaviors are pinpointed with the performance of a specific child on a specific task in mind. But it is equally important to remember that these behaviors are only a part of a total and integrated picture of life. The behaviors you measure should "lead somewhere." By using developmental sequences, or even the structure of a formal task analysis, as a framework for pinpoint generation, you are more likely to identify individual pinpoints without forgetting the broad scope of skill development.

Having a strategy in mind for selecting the pinpoints from developmental sequences and task breakouts makes the process go more smoothly:

— Survey the sequence's major division headings for a quick overview. A brief scan often reveals the chief concepts or skills toward which behaviors are expected to build. Consider, for instance, the headings offered within a small portion of the mathematics sequence in *The Developmental Resource* (Cohen and Gross 1979), shown in Figure 2:2.

— Make a list of the content covered. The sequence in Figure 2:2 provides such a break-out for you, since the headings themselves indicate major concepts and what general order their acquisition follows.

— Select stressed behaviors. Here, some of the division headings indicate specifically the behavior you would call upon (e.g., Numeral Identification Skills—the student *names numerals, writes numerals, matches numerals*). Others specify content but do not specify behaviors. For these, you must check the sequence items themselves. This particular sequence happens to allow a range of appropriate behaviors, which includes saying numbers, writing numerals, placing numbered cards in order, and pointing to appropriate answers.

— Relate individual needs to the sequence or framework you have outlined. Remember, this is only a framework for suggesting behavioral pinpoints. The pinpoints you finally generate must reflect the behavioral strengths—and weaknesses—of the individual pupils with whom you work.

CHOOSING PINPOINTS THAT BEST DESCRIBE PERFORMANCE

Sometimes what initially appears as the easiest pinpoint to select is ultimately not the most sensitive to the kinds of questions you have about the child's performance. Consider the following pinpointing issues in relation to the behaviors you have already selected. Ask yourself whether the pinpoints you have chosen help you focus on the issues most important to you.

Tool Behaviors

Programs may differ markedly in the content areas they cover, and in the complexity of the skills and concepts about which they require the child to develop an understanding. Many, nonethe-

Estimating Group Size

Rote Counting

    1. Digits, 1-10
    2. Digits, beyond 10
    3. Digits, beyond 20

Numeral Identification Skills

    1. Matches or marks numerals
    2. Names numerals
    3. Writes numerals
    4. Seriates numerals

Rational Counting

    1. Early counting trends
    2. Counting moveable sets of objects, 1-5
    3. Counting moveable sets of objects, 6-10
    4. Counting moveable objects, 10 and more
    5. Counting fixed, ordered sets, 1-5
    6. Counting fixed, ordered sets, 6-10
    7. Counting fixed, ordered sets, 10 and more
    8. Identifying set with stated number
    9. Counting out specified subsets
    10. Marking specified number of objects in larger set
    11. Numeral-set correspondence

less, call upon the child to respond to the material and demonstrate understanding, using one or another of a limited number of basic behavior types. That is, they call for oral responses, writing answers, marking or pointing to appropriate choices, and so on. We call these behaviors *basic*, or *tool behaviors*. They fall into three distinct categories: *write, say,* and *do*. Figure 2:3 contains examples of behaviors commonly pinpointed for each of these categories.

These behaviors are crucial, because a child's performance on a task cannot be any better than his ability to perform the tool behaviors it requires. In order to maximize a child's performance capabilities, many teachers who measure child performance find it valuable to give pupils opportunities for extra practice on tool skills, outside the context of the instructional program itself. Such practice may include use of skill sheets, on which the child performs the specified tool behaviors as quickly and accurately as he can for a short amount of time each day.

Some children, for instance, have a more difficult time dealing with an arithmetic problem simply because they have trouble forming the numbers or writing them quickly. Such a child might do considerably better in his program if he were given a separate opportunity each day to practice just *writing numbers* alone. The same procedure has been generalized successfully to other areas (e.g., by giving children the chance to practice in isolation behaviors such as *writing letters, drawing lines,* or *making checkmarks,* without the added requirement of performing them in the context of a program).

Practicing tool behaviors is intended to make production of the tool behavior so easy a part of the child's repertoire that performance patterns truly represent the youngster's grasp of program material, uncomplicated by problems he might have producing the tool behaviors that the material demands. Practice, or drill, of tool behaviors may evoke images of what is commonly called "rote learning" and all the negative connotations that keep this phrase company. In some ways, rote learning is an appropriate label for such tasks; but it should be viewed only as a means of facilitating later, more complex learning activities. When a youngster is capable of producing the basic response easily and rapidly, he is then required to make higher-level distinctions in learning situations in which conceptual elements are the most important.

FIGURE 2:3. *Example Tool Behaviors.*

"Write" Behaviors

- — Writes letters
- — Writes digits
- — Marks x's, check-marks (to indicate correct answer)
- — Circles response
- — Draws line (to match appropriate responses)

"Say" Behaviors

- — Says letters
- — Says sounds
- — Says words
- — Says numbers

"Do" Behaviors

- — Points to letters
- — Points to numbers
- — Presses buttons
- — Turns flashcards
- — Turns pages
- — Matches items

Describing Complex Performance Domains

Not all programs, of course, ask for responses in terms of tool behaviors, nor should they. Tool movements are important, but they cannot possibly provide the entire picture of performance. Some teachers, in fact, feel strongly that performance in certain areas cannot be reduced to behavioral terms at all, and that any attempt to identify measurable behaviors associated with these areas results only in treatment of relatively unimportant surface skills. They might agree that it is acceptable to define behavioral goals for basic skill areas, or for youngsters in special education programs, but would hesitate to use behavioral descriptors in regard to expectations for higher cognitive or affective domains.

Associated with the issue of assigning behaviors to signify performance in complex and intricate subject areas, is the touchy problem of giving behavioral dimensions to complex characteristics of performance such as creativity, talent, or genius. Some of us may state outright that it is patently ridiculous to consider these qualities in a behavioral context. Others may beg the question by pointing out that these terms are normative and subjective, and vulnerable to such bitter argument that they are better left alone. Yet, since we want our children to be able to take off with the skills we give them, to use their learning in divergent, inventive, and imaginative ways, teachers really need to focus upon methods by which they can recognize and evaluate the creative use of skills.

Since creative accomplishment is usually judged on the basis of some kind of performance, it should be possible, with a little extra effort, to construct clear definitions of behavior describing this performance. And why not try to put these more complex performances into terms that can be observed, measured, and communicated? This is your chance to try your hand at creative pinpointing. It helps if you can find a source that has already done some legwork for you.

Pinpointing behaviors in the cognitive domain. Cognitive skills involve the use of what the child perceives or knows. Perhaps among the best known behavioral frameworks for analyzing cognitive skills and constructs, is Benjamin Bloom's *Taxonomy of Educational Objectives in the Cognitive Domain* (1956). In the course of this work, Bloom describes the strategies by which individuals make use of information in terms of six progressively more complex levels. For each level of performance, he includes examples of common educational objectives drawn from the literature, and stated in behavioral terms.

Bloom's *Taxonomy* is, of course, only one way to view how students deal with instructional content. But it is exciting because it suggests that we *can* define behaviors that bring thinking—an internal, and thus, invisible, collection of mental behaviors—into the arena of observable actions.

Since the behaviors described in the *Taxonomy* are generally stated in terms of megachunks, teachers using it will often want to transform these chunks into smaller, more precisely defined performance units which can be readily identified in the student's work and easily explained to the student himself.

For example, Ms. Frazer, who teaches a senior high school introductory child development course, has modified the *Taxonomy* framework to generate a series of pinpoints for evaluating the work of her students. The tasks she constructs call upon students to demonstrate successively more complex thinking skills.

Ms. Frazer adopted a problem-solving approach to instruction, because she is convinced that the development of adequate problem-solving skills is one key to successful parenting. She wants to teach her pupils to analyze situations in terms of the questions they present; then, to identify, sort, and select a viable solution from the options available.

One of the course's most popular instructional units deals with simple behavioral management techniques. Videotapes of youngsters interacting with their parents at a local co-operative preschool are used throughout this unit. In the preschool, parents take an active role in the classroom, sharing management and instructional responsibilities with an assigned teacher. This structure gives an excellent opportunity to study parent-child interactions.

Realizing that most of her students are a long way from her final goals for them, Ms. Frazer has created a graduated series of tasks, each requiring a slightly more difficult level of performance than the last. As Ms. Frazer outlines the unit, she describes the cognitive level she is aiming at, the instructional content the pupils will be dealing with, and the behavior they must perform to demonstrate that they have dealt with that content at the appropriate level. The outline she creates is similar to the one shown in Figure 2:4.

FIGURE 2:4. *Pinpointing Behaviors in the Cognitive Domain.*

Knowledge Level

Definition: The recall of specific and isolable bits of information, e.g., defines technical terms; recalls dates, events, etc.

Lesson Content: The students read text and teacher-made materials describing the nature and use of several common behavior management techniques. Given a list of the techniques they have studied, the students are asked to briefly define each technique.

Behavior: Writes statements defining terms.

Comprehension Level

Definition: Understanding of what is communicated, making use of what is being communicated without necessarily relating it to other material or seeing its fullest implications, e.g., translating information into own words.

Lesson Content: The students are shown videotaped segments of interactions between parents and children at a local cooperative preschool. Each segment demonstrates the use of one of the behavior mangement techniques they have learned about. A narrator describes the techniques seen on the tape. The students are later asked to describe the techniques they saw being implemented in their own words.

Behavior: Writes statements offering descriptions.

Application Level

Definition: Use of abstractions in particular and concrete situations, e.g., comprehends material presented and uses that information for the solution of unrelated problems.

Lesson Content: The students view additional videotaped segments, this time without accompanying narrative. They are asked to decide what behavior management techniques they see being carried out by the parents on the basis of former readings and observations.

Behavior: Writes statements naming techniques and lists points supporting choice.

Analysis Level

Definition: The breakdown of a communication into its constituents or parts so that the relative heirarchy is made clear and/or the relations between the ideas are made explicit, e.g., evaluates relevancy of information, distinguishes between facts and inferences, identifies main idea or selects best answers.

Lesson Content: The students see a series of taped segments illustrating the problem behaviors of one child as he interacts with his parents in the preschool setting. This is accompanied by videotaped discussions between the child's parents and the parent group. In these discussions, the child's parents describe what they feel to be difficulties they have with the management of their youngster. The students are asked to analyse all the segments and to arrive at a behavioral description of what they observe to be the dominant pattern of parent-child interaction, offering justifications for their conclusion.

Behavior: Writes statements containing description and lists supporting points.

Synthesis Level

Definition: Putting together parts to form a whole, arranging them and combining them in such a way as to constitute a pattern or structure not clearly there before, e.g., combining segments of information with other bits of information to come to conclusions not necessarily there before.

Lesson Content: The students view a new videotape showing implementation of a behavior plan worked out between the child's parents and the parent group. After observing how this new plan is carried out and the interaction patterns that result, the students are asked to

Behavior: Writes statements describing new behavior plan, listing selected elements from several management options available.

Evaluation Level

Definition: Judgments about the value of materials and methods for given purposes, e.g., compares a given with specified criteria, discriminating between positive and negative features of the work, describing those features, drawing a conclusion concerning the quality of the item, and justifies that conclusion.

Lesson Content: The students view a new videotape showing implementation of behavior plans worked out between the child's parents and the parent group. After observing how these new plans are carried out and the interaction patterns that result, the students are asked to consider the positive and negative features of the plan as they see them. Further, they are asked to judge the value of the plan based upon such criteria as: (1) the immediate effect they see it having on the child; (2) their perception of the needs of the child, parent, and preschool; and (3) their own conception of an "ideal" plan.

Behavior: Writes statements outlining argument's key issues, listing supporting points for each key issue raised.

Ms. Frazer's plan shows just one way that behaviors demonstrating a variety of cognitive skills could be identified. Although the example concerns students in a high school class, higher cognitive capabilities are by no means limited to this population. They can be observed in one application or another at almost every age level.

The teacher of a third-grade class might analyze the ability of his pupils to *apply* basic computation skills in order to solve story problems. He could look at a behavior such as *writes digits* as significant of the mental process involved; or, he might even ask his students to "talk through" each problem they solved, recording their thought processes on a tape recorder. The teacher could then listen to the tape at a later, more convenient time and analyze a behavior such as *verbalizes logical steps*, concentrating on the process itself, rather than its product.

The behaviors you choose to demonstrate cognitive abilities are virtually unlimited. Ms. Frazer used the same core behavior for each level of thinking, and her efforts are an example of how the behavior can stay the same although the level of complexity increases. But, she might have set up her exercises so that different behaviors were required in each instance.

For example, she might require *writes definitions* at the knowledge level, *writes descriptions* at the comprehension level, and then call on entirely different behaviors, such as *provides actual demonstrations* of a possible plan, at the synthesis level.

Although the behaviors chosen for these exercises were limited to writing, Ms. Frazer might have mixed *write, say,* and *do* responses, allowing students a variety of options for demonstrating what they knew.

As we pointed out at the beginning of this discussion, Bloom's is not the only framework available for analyzing higher levels of cognitive functioning. A teacher of creative writing, for example, might decide to use any one of many fine composition and style manuals available, formulating pinpoints based upon the rules and strategies found there. Strunk and White's *The Elements of Style* (1979) is an excellent example of just such a manual, easily adapted to pinpointing efforts. Its rules and suggestions make not only for enjoyable and informative reading, and excellent writing; they can also be quickly converted into simple, concise behavioral pinpoints.

The book's final chapter, "An Approach to Style," describes behavior which is absolutely essential to good writing. Style is a quality so elusive and highly personal, that even the chapter's author admits, "Here we leave solid ground. Who can confidently say what ignites a certain combination of words, causing them to explode in the mind?" Yet E. B. White's "suggestions and cautionary hints" provide a solid framework for pinpointing at least the rudiments of writing style.

Martin Johnson, a teacher who bases his instruction on the Strunk and White book, uses the following behavior to signify just one of the components he considers crucial to the development of a good writing style: He encourages his students to *use simple, direct language.* Since what he has identified is a behavior chunk, the teacher decides to define very specifically those elements he considers to be appropriate and inappropriate demonstrations of the pinpointed behavior. Mr. Johnson makes a list of correct and incorrect response categories, which he keeps on the board to remind students of some stylistic "do's" and "don'ts." Figure 2.5 illustrates his definition.

This instructor has taken most of his reminders from the Strunk and White guidelines. They are easily understood by the students, and easily confirmed in the body of whatever work the teacher is correcting. Mr. Johnson agrees with the "Style" chapter's author that these behaviors describe the crucial differences not so much between writing that is right or wrong, but between those efforts that are distinguished and those that are not. This is an extremely fine distinction, but it can be made in simple, clear behavioral terms.

Behavior in the affective domain. The affective domain concerns a child's emotional make-up, evidenced in such things as *interests, attitudes, values, likes, dislikes,* and *social adjustment.* You will immediately recognize *interests, attitudes, values,* and all the rest as behavior chunks of the largest magnitude. These are internal behaviors and, as such, are not usually thought of as observable. We make the assumption, however, that certain behaviors are directly related to an individual's interlocking feelings and emotions.

Some of these observable behaviors are very subtle—a raised eyebrow, or pursed lip—and may occur only fleetingly in the context of a specific, isolated situation which triggers a particular emotional response. Others may be more overt and persistent, consistently cropping up across a variety of situations which require a particular kind of social adaptation.

If we can identify these behaviors and the emotional context in which we would expect to find them, we may be able to help youngsters who are considered to have problems with "emotional adjustment." We can enable these youngsters to identify, relate, and even change the way they behave in interactions with parents, peers, and other concerned individuals. A variety of books,

FIGURE 2:5. *Pinpointing Some Elements of Writing Style.*

Behavior: Use Simple, Direct Language

Try always to:

1. Choose simple words over the fancy and pretentious;
2. Use dialect sparingly, and then correctly and consistently;
3. Use English rather than foreign words (including Latin, for the youngsters in the class-rooms who may be suffering proudly through Latin III); and
4. Choose the standard word or phrase (e.g., theft) over the colloquial and offbeat (e.g., rip-off).

Try never to:

1. Make choices in the line of "tummy" over "stomach," "beauteous" over "beautiful;"
2. Use dialect extensively, inconsistently, or inaccurately;
3. Sprinkle work liberally with foreign expressions;
4. Use the current idiom excessively (e.g., uptight, rap, dude, gross, cool); and
5. Make extensive use of incorrectly "verbized" nouns (e.g., accessorize, prioritize).

tests, and taxonomies may serve as frameworks from which teachers can extract lists and descriptions of significant affective behaviors.

The following are only a few of the strategies a teacher might employ to pinpoint and work with behaviors that make up the fiber of the child's emotional and attitudinal response.

— At a very simple level, a teacher who suspects that one of her pupils suffers from poor self-concept might try to identify behaviors that indicate this problem (e.g., *hitting* on the playground). At the least, she can work to reduce the incidents of hitting, making the child more tolerable to be around. In addition, she might identify and work to increase behaviors that would help him to become "a better friend," or "more popular with his peers," thus allowing him actively to improve his battered self-esteem.

— Another child might be encouraged to "explore his feelings" about himself by *circling* "happy" or "sad" faces at various times throughout the day (e.g., after completing an assignment, engaging in a game or activity with other children, or in response to questions or situations posed by the teacher).

— Parents can investigate "negative feelings" about their child by *making a list* throughout the day, describing the things the child *does* which either annoy or please them. Such an activity draws parents into pinpointing for themselves "trouble" behaviors that they would like to work on, whether these are behaviors that they would want the child to do more of, or to do less frequently. Sometimes a parent who is convinced his child never does anything right is surprised to find that his child doesn't do nearly as many things to "bug" him as he had thought. In fact, the child may do many more positive than negative things; but, whatever it was that bothered the parent, overshadowed the child's positive behavior. Other parents might be encouraged to keep only a "positive" list, forcing them to focus on the things they like about what their child does.

— Teachers eager to help youngsters develop social skills and attributes important to being at ease in social situations, might begin by imagining someone who possesses the desired attributes, and listing the *behaviors* that person would exhibit in a variety of contexts (e.g., uses "open body language": smiling at others, making eye contact, touching others, etc.). The teacher could then imagine an individual who possessed none of these attributes and list the behaviors that individual would be likely to exhibit in the same situations (e.g., "closed body language": averting gaze, mumbling, etc.).

The two lists represent pinpoints that a teacher could observe and work to promote in the course of the normal school day. Structuring a variety of "mock" situations requiring practice of such behaviors (e.g., job interview or playground scene, depending on the youngster's level), would allow students the opportunity to identify and practice (role play) these behaviors for themselves.

None of these suggestions need be particularly time consuming to carry out, and teachers may find a number of eager helpers among students and parents as they search for pinpoints relevant to the development and polishing of social skills.

Fair Pairs

You've heard the old saying, "There are two sides to every issue." Behavioral issues are not exempt. Depending on the behavior you have in mind, you may well find that you need to examine both a correct and an error component, or even two completely different behaviors—one positive and one negative—both of them part of a single behavioral picture.

Oral reading is an example of an academic response frequently described in terms of both correct and error elements. Popular commercial reading inventories, such as Silvaroli's CRI, define both correct and error responses: *corrects* are word responses consistent with the visual cue (the printed word); *errors*, on the other hand, include *repetitions, insertions, substitutions, omissions,* and *asking for assistance.*

Defining and examining error responses for measurement purposes is useful in that it provides a total picture of performance. Teachers hold a variety of beliefs about how correct and error patterns in reading relate to one another. Some assume that, if you pay attention only to what is correct, errors will automatically disappear. Others feel that, when correction of errors alone is emphasized, correct responses eventually increase. A third group maintains that correct and error responses must each receive attention. Regardless of which theory you favor, you will need to look at both components to determine whether or not your assumptions are proving correct.

For the teacher who does believe in "treating" errors, the operation involved in breaking out error components allows the teacher to perform a quick error analysis, determining exactly what kinds of errors the child makes most commonly, and whether some should be given different consequences than others.

"Fair pairs" work for social behaviors, too. At the same time you pinpoint a negative behavior that you would like to see eliminated, it is equally important to specify at least one positive behavior to build in its place. It's so easy to say, "OK, I'm going to get rid of throwing, swearing, and hitting, and then that kid is going to be a different kind of human being!" But different, *how*? Simply removing problem behaviors does not guarantee that a child will replace them with more desirable forms of response. Sometimes he has no knowledge of what it's like to perform a positive behavior. If you decide to help a child decrease *hitting*, what more desirable response can you work on putting into its place—*sharing toys, making appropriate conversation, pitching baseballs*? Call it "building a positive self-image" or anything you want; just make sure you examine both sides of the behavioral coin.

Response Modes

By *response mode*, we mean the particular expressive form a response takes—whether it is written, spoken, or whatever. In most learning situations, it is possible to identify numerous ways that a child can demonstrate a particular skill. In math, for instance, the answers to basic facts can be written, spoken, circled, or pointed-to; you have a variety of options. A child's success in the program may depend upon what ease or difficulty he has in performing the movement cycle called for.

It often surprises teachers to find that a student, who seems incapable of providing a correct response within a given program, is experiencing difficulty, not in learning the required material, but rather, in performing the specific movement cycle (or tool behavior) that has been designated to demonstrate learning.

Johnny, a pupil in a resource classroom, has been toiling with little success at simple addition facts for over two weeks. One day he collapses in tears at his desk, and is overheard to wail, "But I *know* the answers—I just can't *write* them!" His teacher questions him and discovers that, indeed, he can *say* all the answers correctly. From that point, she asks Johnny to read the problems and their answers into a tape recorder or aloud to her, while he continues on the side to master written number responses.

The issue of response modality is an especially important one within programs for the handicapped, who are likely to demonstrate a variety of response deficits. While instructional materials often assume that all students working in them will be capable of responding in a stipulated way, most programs can be adapted to account for individual response needs. As a rule of thumb, if there is more than one way that a child can demonstrate a skill, select the behavior that is both closely related to the skill in question, and, at the same time, most favorable to the child (that is, takes into account his own special abilities or disabilities).

Reflecting Learning across a Wide Range of Developmental or Subject Areas

Many of us suffer from pinpointing "blind spots." That is, we concentrate most ardently on those pinpoints that stick out as "trouble makers" in the classroom—behaviors falling into the category of civil and uncivil disobedience—or upon those which are a part of our own favorite instructional areas, and in which we ourselves demonstrate the most expertise. The third-grade teacher who loves reading, who has taken over thirty hours of extra course work in reading instruction, and who would rather die than run a quarter mile, cannot be blamed for choosing to put extra effort into his reading program, and staying as far away from the soccer field as possible. Yet, it is extremely important to recognize the diversity of behaviors a youngster must master and demonstrate as a part of coping with everyday life. Broader pinpointing capabilities are essential to the teacher who is concerned with selecting learning activities that reflect the "whole" or "total" child.

By using a framework such as a developmental scale or sequence, which calls your attention to the wide range of developmental areas important to a complete picture of growth (e.g., motor, social, language, cognitive), you are certain not to overlook skills that might otherwise fail to come under scrutiny.

Accounting for All Stages of Learning

Just as "thinking" may be demonstrated at different levels of complexity (e.g., simple fact recitation versus application of known facts to an unfamiliar situation), other skills may be demonstrated at different degrees or stages of proficiency. Some authorities have even tried to identify "levels of learning" or a "learning hierarchy," suggesting that a child's performance of a skill varies according to the level—acquisition, mastery, maintenance, or generalization—at which he performs. The stages of such a hierarchy are still a matter of speculation and need considerable data for confirmation; but you have, no doubt, observed differences in the behavior of a child painfully combining the skills required to pull on a jacket or coat for the first time, with his behavior a few months later, when he can put on that jacket or any other similar article of clothing, as well as dress a favorite doll in a jacket, sweater, or coat.

If you think that a program you are considering does suggest hypothetical stages of learning, consider these issues:

— Can you identify the terminal behavior of the program?
— Is the terminal behavior identical to the initial behavior you require? That is, have you selected a pinpoint which can be tracked and compared across a variety of stages of learning?
— Will different behaviors be required for various phases of learning? For example, in the early stages of a time-telling program, *identifying numerals to 12* may be an important

behavior. As the program progresses, however, you find yourself needing to pinpoint and measure other behaviors, including *identifying the hour and minute hands; reading time to the hour; telling the half hour and quarter hour; reading time in five-minute increments;* and *reading time in minute increments to sixty.* As the child nears mastery, and learns to generalize time-telling skills to outside situations in which they will be important, still other pinpoints are identified, such as reading the clock and figuring if it's time for recess, when it is time for dinner, or how long before his favorite TV program.

REFINING BEHAVIORAL PINPOINTS

One final shopping exercise awaits you. Now that you've pulled your pinpoints off the shelf, you have to decide how well you really like the product in your hand. Does it look the way you expected it to? Is the packaging right? Is what you're getting of a convenient size and amount? And, most important, can you tell the contents by the label?

By this time, you have a better notion of your own priorities in selecting behavioral targets, as well as of the sources and resources that will guide the pinpointing process. What remains is a final check of the pinpoints you have identified to make sure that they will do the best possible job of representing both program goals and individual child needs. Suppose you put a pinpoint through this final evaluation process and find that you are not, after all, satisfied with the way it describes the problem. What do you do? You can trade it for a completely different behavior. You can decide to forget the whole thing and walk away empty handed. Or, you can try to modify the pinpoint in a way that will make it acceptable.

We call this process *refining* behavioral pinpoints. Maybe you're convinced that you've found the "perfect" suit, if only you can get the tailor to take it in a little here, shorten it a little there. Often some judicious pruning and polishing is all that's required to make a pinpoint fit your needs exactly. The following are important issues that should help you decide if your pinpoints need refining, and figure out how to trim them down to size.

Establishing Uniform Pinpoints

Whenever possible, select a standard pinpoint that will apply appropriately throughout a particular text. Frequently, materials call for a variety of quite different types of performance. An elementary math workbook, for example, may present a number of concepts, including simple addition and subtraction, change-making, and borrowing and carrying. It asks the child to demonstrate understanding of these concepts in numerous ways: by *marking answers, writing numerals, saying answers, pointing to correct alternatives, drawing lines to match sets, drawing or coloring objects to illustrate concepts,* even by *cutting out and pasting forms.*

A child's performance seldom stays exactly the same from day to day. Some of the differences you see—increases or decreases in speed or accuracy—may relate to the child's changing understanding of the concepts being presented. But it is also possible that some performance changes can be attributed as much to differences in ability to perform certain responses (*say, write, draw, writes digits, draws circles,* or *says numbers*) as to changes in actual learning.

One way to get around this problem is to try to zero in on a single pinpoint that could be used appropriately throughout the entire instructional text. Pinpointing a movement cycle that will remain uniform enables you to track performance across time and to compare the student's abilities at program entry to his performance as new skills are introduced and the material becomes increasingly complex.

Picking a uniform pinpoint is a strategy that also allows the teacher to compare a child's performance across a variety of skill areas, or to compare learning ease under one instructional approach as opposed to another. For example, you can select a single movement cycle such as *reads words orally*, and use it to determine which of two different reading approaches is most effective with the child. By selecting stories of equivalent readability from one reader using a linguistic approach and from another employing a sight word approach, you can compare the child's response in one instructional method against another. If you introduce still another type of task—for instance, a graded word list—you can easily determine how variables such as context affect the child's performance.

Although the examples we have used here are limited to the instructional areas of reading and arithmetic, it is easy to see how selecting a single, standard unit of behavior as a measure of learning allows you to contrast, with ease and accuracy, a youngster's performance from one concept to another, from one book in a series to the next, across subject areas or across instructional approaches.

Functional Definitions

Just as we assigned behavior chunks functional definitions (identifying criteria concerning space, time, and effect), we may sometimes find it necessary to apply functional definitions to movement cycles. A teacher interested in decreasing the amount of time a youngster in her class spends dawdling before returning to his seat after recess and lunch might define the problem this way:

> She wants to increase the number of times each day the child *returns to seat* (on time). She defines returns-to-seat functionally, establishing a time criterion by specifying that, for a response to be correct, the child has to take his seat *no more than three minutes* after the bell is rung. In this way, the teacher is absolutely certain that what she counts as correct or error remains consistent from one time to the next. She can set definite limits much more easily for the "dawdler." In addition, she can explain her pinpoint, and the program she eventually creates to deal with it, more clearly to other individuals who also find the amount of time it takes for this child to "settle down" a problem.

Behavior Slicing

Teachers often put considerable effort into formulating an instructional program aimed at improving the pinpointed behavior, and sometimes even into measuring performance, only to find that the behavior itself has somehow eluded them. They did not define it quite precisely enough, and the behavior they came up with is not sensitive to the changes they should be seeing; or maybe they got overzealous, and refined their definition to the point where the tiny piece of behavior they're looking at simply isn't sufficient to cover the broader range of performance the child demonstrates.

Pinpointing behavior is like being given a cake and told to help yourself. You can cut any size slice you want, from a wonderful fat wedge (a behavior chunk) to a paper-thin sliver (a movement cycle). The important thing is, to cut a piece the right size to meet your needs. Cut too big,

Slicing Behavior: A Piece of Cake

and you find yourself having to put some back; slice too fine, and you end up coming back for more.

In oral reading, for example, pinpoints are available in a number of sizes, from very gross to very fine: *reads books, reads paragraphs, reads sentences, reads phrases, reads words, reads syllables,* and *reads phonemes.* You could, of course, use *reads books* as your movement cycle, and count the number of books a child read a semester as a measure of performance. But the pinpoint is so broad that it is not at all sensitive to daily performance changes. The same problems hold true with pinpoints such as *reads pages, paragraphs,* and *sentences.*

While teachers frequently assign a child to read a designated number of pages, paragraphs, or sentences before the next day's class, these pinpoints are not really well suited to accurate analysis of progress in reading. Part of the problem lies in the inconsistent size of the pinpoints, making comparisons of performance all but impossible. Size varies immensely from one sentence to another. To say that a child reads ten sentences correctly one day, and twenty correct on the next, may make it sound as if the child's reading is improving; but perhaps the sentences of the first day are much longer. And again, what counts as an incorrect sentence? Is there any difference in a sentence read with one error and one read with seven errors? How does the pinpoint account for these differences?

When you find a behavior that is too big to manage, *slice back.* In the case of reading, slicing the behavior back to the size of *reads words* will result in a pinpoint that is (1) of fairly consistent size, and (2) sensitive to performance variables (i.e., you can easily define correct and error responses, and can even analyze error types, giving you a much more detailed picture of the child's reading patterns). Sometimes you will identify a performance problem that indicates still further slicing is necessary. For example, if a beginning reader consistently demonstrates difficulty discriminating the differences between phonemes such as *a, o,* and *u,* you might decide to give him extra practice on drill sheets containing random vowels. The movement cycle would now be sliced back to *reads phonemes.*

It is possible to go in the other direction, and slice too finely. If you decide to analyze the child's total reading performance with a pinpoint such as *reads phonemes orally,* you are likely to be spending all day just counting the behavior. Also, for most purposes, looking at the phonemes a child produces would not provide the most functional picture of his reading performance in context.

Before you begin planning, initiating, or changing a program, you might do well to consider Goldilocks' persistence, and keep hunting for a pinpoint until you find one that is just right! Whenever you are trying to decide whether a pinpoint needs slicing, and how fine to slice, consider these suggestions:

— Each repetition of the behavior should represent essentially the same amount of behavior. This makes for much more consistent measurement and comparison of behavior from time to time, should you be interested.
— The behavior should be directly and meaningfully related to the overall instructional objective.
— The behavior should be easy to count, making whatever measurement techniques you employ practical and efficient.

CONCLUSION

As you finish reading this chapter, you may be asking yourself, "Do I have to consider *all* of these issues *every time* I decide to pinpoint a behavior?" Of course you do not. But we hope that you will consider at least some of them *before* you begin the pinpointing process. Otherwise, you may learn, as we did, that these are the questions you come back to again and again.

The information summarized here was gathered first hand, through long, often hard, experience. It's all too easy to labor over a particular skill for an entire school year, only to have someone come in at the end, look at the program, and say, "So what?" You may find that the behavior you pinpointed isn't really relevant to the curriculum as a whole, that it fails to account for important levels of complexity, or that it is not, afterall, of top priority for the child. All learning is a series of advances and retreats, and you will no doubt experience many pinpointing retreats at the start. By considering the questions and issues raised here, however, you reduce your risks considerably. At the same time, you increase your chances of identifying behaviors that are relevant both to the curriculum and to the child's most pressing needs.

Assessment: Targeting Behaviors for Intervention

Chapter 3
Assessment as a Basis for Pinpointing

Assessment is among the most popular of strategies for determining behavioral pinpoints. The assessment process offers an organized, directed approach to identifying each child's individual performance strengths and weaknesses. Specifically, it allows the teacher to establish the child's present level of performance, to determine where the child should be placed in an instructional program, and to formulate ideas about how such a program can best be carried out.

Teachers who decide to devote time to initial evaluation often find that an overwhelming number of testing options are available to them, for almost any skill area they want to assess. They can purchase elaborate, expensive, commercially developed kits, complete with manuals, recording forms, and special testing materials, or they can create and implement their own "home grown" assessment tools. They can administer any number of formal, standardized measures of performance or employ informal inventories, criterion tests, and probes. A teacher may be convinced of the need to assess, and yet experience considerable confusion trying to decide what specific kind of assessment tool will provide the most usable information.

Much of this confusion can be avoided, if you think of assessment as merely a *question-asking process.* All that any assessment tool is—no matter how fancy, or how unassuming—is an *organized question-asking device.* The basic differences among testing tools lie in the way they formulate and ask questions about performance.

Formal, standardized tests tend to pose questions that are related only indirectly to specific instructional issues. They may reveal a general learning deficit (e.g., that a fifth grader reads at a 2.5 grade level); but they do not usually pinpoint specific skill deficits that clue the teacher in to immediate possibilities for remediation (e.g., problems with dipthongs or consonant blends). Informal assessment, conversely, focuses much more closely on the individual issues of program placement and development.

This chapter is intended to give you a closer look at the differences in the types of questions asked from one kind of assessment instrument to another.

USING FORMAL AND INFORMAL TESTS TO IDENTIFY INSTRUCTIONAL PINPOINTS

Formal Tests

Formal, or standardized tests are so called because they detail a standard set of procedures to be strictly adhered to in test administration. Since these procedures are a part of the conditions under which testing norms are established, they must be followed without deviation: altering of testing procedures to suit the individual invalidates test results.

Not only must each test be administered under a rigid protocal, but each is also generally best suited to only a single, narrowly defined purpose, such as screening, classification, or diagnosis. For example, a screening battery is designed to sample behavior in relation to a broad range of pinpoints in a small amount of time. The major purpose of screening is to determine only whether or not a child needs further evaluation. It cannot, however, provide the kind of in-depth examination of a particular learning problem that would be possible in a diagnostic examination.

For the most part, the questions standardized tests ask bear only a very indirect relationship to what we do on an ongoing, day to day basis in the classroom. While their results can clue us in to certain performance discrepancies, especially in relation to the child's peers, they don't answer questions of the nature, "What in the world am I going to do for the kid's reading program on Monday?" If you intend to review such test results with the purpose of drawing programming pinpoints from them, you should keep in mind a number of questions about the construction of the tests you use.

Does the test discriminate unfairly against the child? Formal, standardized tests are *norm referenced.* That is, they purport to compare a child's performance with that of a *like* group of individuals.

Unfortunately, the populations on which these norms have been based are not always representative of the youngsters who are eventually tested. By ignoring differences that can exist in certain subgroups, these tests leave themselves open to considerable testing bias, most markedly, bias that is culture- or language-related. If test items represent a knowledge that a child does not have of the culture, or if the test is given in a language that is not the child's first language, low performance may result. While the child might know the concepts tested, or understand culturally equivalent concepts, he is discriminated against at the onset by assumptions inherent in the test construction itself.

Bias also arises in regard to certain handicapping conditions. For instance, test presentation may be primarily visual, discriminating against the child who exhibits some visual impairment, or who simply performs better under auditory presentation. Perhaps the test requires written answers, greatly handicapping the youngster who knows the answers, but performs at a disadvantage on any pencil-and-paper task.

Is the test reliable? In the material accompanying a standardized test, you will usually find a stated reliability coefficient. This figure is important because it provides the basis for what is known as the "standard error of measurement," a statistic describing possible variations in scores of a single individual if he or she were to take the test a number of times or took alternative forms of the test. The standard error represents an estimate of the amount of error which might typically accompany the child's true score and gives information about the confidence with which a score can be interpreted. No score can be interpreted as a discrete point, but rather as a range with a band of probable inaccuracy on either side of the obtained score. The larger the standard error, the less certain you can be of the true score.

Is the test valid? Whether or not a test measures adequately what it claims to measure is another matter to consider as you contemplate using test results to plan instructional programs. We will cover neither reliability nor validity in depth here, but urge you to consider both these factors before applying standardized evaluation procedures in your classroom.

How far can test scores be applied directly to curriculum planning? Reliable and valid standardized test scores are most likely to give you a broad picture of the child's abilities in relation to general skill areas, or in relation to the performance of other children considered to be in a comparable "norm" population. But the tests are generally not constructed to give you precise information about individual strengths and weaknesses (e.g., rather than having attained a 2.0 level in reading, are the child's sight-word recognition and word-attack skills vastly different? Is a particular math score related to difficulties with addition, subtraction, time telling or all three?).

Because standardized tests seldom relate information directly to discrete curriculum areas, you are likely to have difficulty determining on the basis of the test results how a child will perform within the specific instructional areas you are planning. Nor can you determine, from these scores alone, whether extra concentration in certain areas is merited.

Do response patterns occur that cue you to look further at presentation and response modalities? Do you see predominant response mode preferences (e.g., does the child do consistently better on those questions requiring verbal rather than written responses?). Does the child favor certain modes of material presentation (e.g., missing all the items in which input modes are visual, but responding correctly when auditory or a combination of visual and auditory input is used)? Your school psychologist or school counselor, trained to interpret test results and, therefore, an excellent resource person to consult as you attempt to identify pinpoints, should be able to help you identify such modality-preference patterns. You may want to explore such response differences further, creating tests to examine specifically the suggested differences, or incorporating these observations in program plans.

Certainly any test created to demonstrate the child's concept understanding, if the results are to be unbiased by response or presentation deficits, would have to keep the possibility of such differences in mind.

Could the testing procedure have been altered to help the child perform differently? In order for standardized tests to get valid results, the protocol or procedure for administering the tests must also be standard. Such tests, therefore, cannot allow much leeway and still guarantee valid result interpretation. For some children, however, you may be able to identify factors in the way the test is administered which, if you can change them, will alter the child's performance significantly (e.g., can some parts of presentation be changed so that particular cues are included; or can alternate ways of responding be allowed; is the examiner a stranger or someone familiar to the child; is the test administered to a group or on a 1:1 basis; does the evaluation take place in a new or in a familiar setting?).

Standardized tests are not, on the whole, designed to answer specific programming questions. The information they provide can, however, be used to identify useful behavioral pinpoints if you keep the issues that have been raised here in mind.

Informal Tests

Unlike the formal procedures just discussed, which bear, for the most part, only an indirect relationship to what you would do on a day-to-day basis, informal assessment procedures have direct relevance for pinpointing and programming concerns. Further, because they do not require standardized conditions for administration, you are able to manipulate certain elements, such as methods for presenting a question or means by which the student may offer a response. Flexibility to explore such factors or variables that the teacher considers potentially important to a child's performance is one of the distinct advantages of an informal measure.

How applicable informal tests are, of course, is a direct function of the care you demonstrate in selecting or creating the test you will use. Consider the following issues:

Deciding where curriculum guidelines are coming from. Most informal tests are set up with the idea of directly measuring performance in relation to objectives in the chosen curriculum. Implementing these tests assumes that you have firmly identified such an instructional sequence. If a complete curriculum sequence is not a part of administrative guidelines given you, and you are in some doubt about where you might start in defining one for yourself, consider:

1. The scope and sequence charts accompanying the commercial instructional materials you will be using;
2. Formal curriculum guides, available through many school districts; or
3. Behavior checklists, available in great abundance (over 200 such lists are evaluated in a work by Walls, Werner, Bacon, and Zane [1977]).

Making sure objectives are clearly formulated behavioral statements. Once you have decided what curriculum guidelines you will be using, analyze carefully each of the instructional objectives that have been specified. Have goals been broken down into series of smaller objectives describing the steps which are involved in the attainment of that goal? Just how carefully have behaviors been defined? Are requirements presented in vague terms, such as "the child should 'understand' or 'appreciate'," without any attempt to specify the precise behavioral indicators involved?

Making certain that test results represent reliable performance characteristics. Plans for remediation of a skill based upon test results are justifiable only if you can demonstrate that the errors seen represent a reliable performance characteristic. It is crucial to be able to recognize the difference between a random mistake and a skill deficit which appears consistently. When testing is based on an analysis of a student's work over a period of time, or is conducted on more than a one-shot approach, this problem is usually avoided. The number of items placed in a test to examine a particular skill may also be relevant. For example, an error on a particular test item which is the only one of its kind on the test, may be simply a random mistake, rather than indicative of difficulties with *all* problems of that type. A very short test given just once will yield far less reliable results than informal tests providing more opportunities for a student to confront certain problematic requirements, and, thus, the teacher more opportunities to observe performance.

Making your tests reusable. If you decide to create your own informal tests or inventories, you are likely to invest a considerable amount of time and effort. Spend that effort well, by making the tests you develop a permanent, reusable part of your classroom materials.

— If you are planning to write a series of tests, develop some sort of master plan for coding these tools to the skill hierarchy or scope and sequence chart you are using as the basis for your program. If you plan to pool efforts with colleagues, you can all reap the benefits of each others' work by organizing tests according to a master plan to which you all agree.
— Plan the number and types of items you will include on each test. You may want to make an inventory comprehensive enough so that each time it is used, only those sections which apply to a given child would be selected and pulled out to be administered.
— Set up a test file, with a system for storing and retrieving entire tests or individual test items.
— Finally, decide what will be the easiest and least expensive method for reproducing informal assessment tools—whether by stencil, ditto, photocopy, or spirit master.

USING INFORMAL INVENTORIES TO ASSESS PERFORMANCE

An informal inventory is designed to provide a basis for developing formal instructional objectives and individual program plans. That is, a formal statement of objectives is constructed based upon the student's performance on the inventory. This sequence of events is in direct contrast to that observed with criterion tests, which will be discussed later, and the distinction will be an important one.

Inventories are generally comprehensive in nature, presenting skills within a hierarchical sequence. They typically include a large number of test items associated with each skill within the hierarchy. The advantage of the informal inventory is that because it is so comprehensive, it is possible to survey child performance over the entire range of skills intended as the subject of classroom instruction. Test results allow you to determine where, in general, the child performs in relation to the total content.

Creating Your Own Informal Inventories

In certain cases, you may find commercial inventory materials useful. Some can be administered as is, while others may be adapted with slight modifications. Many, however, will not be as comprehensive or directly applicable as you would like. These can still be used to offer initial guidelines for constructing your own set of informal testing tools.

Identifying a sequence of necessary or desired skills. The first step in developing a skills inventory is to establish the sequence of skills over which performance will be tested. Items can be created to survey performance of specific skills listed in behavior sequences and checklists. To determine what textbook, or what part of a textbook in which a student should begin work, you may want to base items on instructional materials actually in use in your classroom. For example, an informal reading inventory can be created by extracting oral reading samples from several levels of the adopted reading text. The child is asked to read selections aloud and answer comprehension questions based upon the selections.

Attempt to confirm where test items, and texts you have taken them from, fit within a broader curriculum perspective. Check whether the publisher has placed concepts at appropriate grade and difficulty levels. Such a check can be performed on graded readers, for instance, by applying a readability formula. For other subject areas, lacking such formulas, text samples can be compared against the scope and sequence charts accompanying other widely used texts, against a standard curriculum guide, or in terms of the recommendations of a local textbook review board. In addition to helping you establish the difficulty level of your adopted text, such efforts allow you to organize overall objectives.

Establishing relevant item context. Skill inventories are not confined to pencil and paper tasks. Whenever possible, the materials which sample performance should be those with which the child will be expected to interact in the "natural" setting. That is, reading and math skills are best sampled using, as described earlier, the actual texts in which you expect daily performance. In a self-help inventory testing shoe-tying or shirt buttoning, give the youngster the opportunity to work with articles of his own clothing. Feeding skills are most realistically examined in the context of a meal or snack. Change-making skills are more accurately assessed using actual coins and bills. All of these settings, events, and materials are closest to the natural environment in which the child is called upon to function; performance in them allows the most realistic and accurate assessment of his capabilities and needs.

Selecting inventory items. After deciding the general context for constructing the inventory, it is important to set some guidelines as to how inventory items will be selected from prepared instructional material.

Decide how many items you need. Select items representing each major topic covered in a skill inventory or in the text you intend to use. For instance, an informal reading inventory is generally constructed by selecting three passages for oral, silent, and listening capacity at each level. If word lists are used, they are constructed based on these paragraphs.

Decide on the length of test items, determining what would be a reasonable length to allow adequate sampling of the child's ability. If you decide to time reading samples, for example, you need to determine how many words will provide enough material to fill the selected sample time (e.g., one minute) at each graded level. If, rather than using a set timing, you want to establish a fixed number of words for each selection, determine what size passage will be representative for a given difficulty level (e.g., 30 words in each passage for preprimer, primer; grades one through four, 100 words, etc.).

Decide on the number and types of comprehension questions that will be required, where such questions are appropriate. If possible, and appropriate, include items that survey a range of thought complexity levels (e.g., requiring skills such as definition of a word in the context of the passage; immediate recall of a fact; drawing an inference from facts presented in the selection; statement of the main idea or conclusion to be drawn from the passage; a listing of the sequences of events and a prediction of outcomes might be included).

As you select items, always look for those which would require complete mastery of the skills in question to complete. Ask yourself, "If the child could do this problem or group of problems correctly at the end of my instruction, would I be satisfied that he had acquired the skill?" If you end up qualifying your answer with, "Yes, but he should also be able to . . . ," you need to reexamine the items you have included.

Deciding where to start sampling. The objective in administering an inventory is to establish at what point in the curriculum the child begins to need extra assistance. It is at this point that classroom instruction should begin. Given the sequential nature of inventory construction, test items are generally easiest at the beginning, progressing gradually to more difficult skills. Since presenting an entire skill sequence for a given area may take more of your time and the child's than you are willing to give, you need to establish rules or strategies stating where sampling begins and ends.

One such strategy involves a "best estimate" method, starting where reports from a referring teacher or where your own informal observation leads you to believe the child is functioning, and continuing to test upward until you are certain that you have determined an instructional "starting point."

Still another strategy involves starting with the most complex behaviors and working backward until you find a place where the child performs comfortably. Instruction might then begin a step ahead of this level, or you might choose to start at this "comfortable" level to ensure initial success. Setting up one level of a primary level math inventory, for instance, one might assume that the most complex task level requires solution of verbal problems stated in paragraph form, involving several steps. If the child cannot deal with these in their original form, you might try verbal inquiry, attempting to discover student strategies in dealing with the problem. Failure at this level would indicate next presenting open mathematical sentences describing the operations involved as computation problems. Further failure would suggest presenting the problem in computation form.

Reading inventories traditionally present a word list before the oral paragraph and its related comprehension questions to determine entry level for oral paragraph reading; however, in the interest of saving time, you might elect to present a list of words in isolation only when paragraph reading falls below instructional level or word-recognition criteria which you have established.

Deciding where to stop sampling. What kind of performance indicates that "this is where testing should stop and instruction begin"? It differs from one instructional area to another, and you may have to consult texts or experts in the field to determine exactly what kind of performance continues to be acceptable and what is not. Of course, total failure to perform is an irrefutable signal; but other, more subtle signals may be observed before the child reaches the failure point.

In reading, for example, three levels of performance are clearly defined: *independent, instructional,* and *frustration.* The *instructional level*—where reading instruction would commence—is the highest level at which a child reads satisfactorily, provided he receives teacher preparation and supervision. Instructional level is frequently cited at 95 percent average accuracy in word recognition and with 75 percent average comprehension. *Frustration*, on the other hand, where the child's reading skills break down, word recognition averages 90 percent or lower and comprehension averages 50 percent or below. The terms "independent," "instructional," and "frustration" may be helpful outside the area of reading, if you can decide upon criteria for these different performance levels as they might apply, for instance, in math, science, social studies, or even self-help.

When inventories are based upon developmental checklists or other forms of behavioral sequences, sampling can almost always be accomplished most easily by beginning with the last items in the sequence first, working backward as the child's responses indicate. It is assumed that most of the earlier behaviors are encompassed in successful performance of the final behavior in the sequence. This behavior is usually the final outcome and the desired performance goal.

One such assessment tool, called the Pinpoint Scanning Device—created to test behaviors making up sequences in *The Developmental Resource* (Cohen and Gross 1979)—implements this most-to-least-complex approach. This particular tool is made up of a number of discrete skill sequences. In order to work through the sequences as quickly as possible, the teacher is encouraged to start at the end of each sequence. Mastery is assumed whenever a child immediately demonstrates consecutive "yes's" on the last three items, and testing of that sequence is discontinued. On other sequences, the teacher works backward to the point where three consecutive "no's" occur, and then begins testing forward from the first skill in the sequence, until the child again begins demonstrating difficulty.

Setting up a test protocol. Try to specify some of the major factors you think might affect this child's performance and take these into account as you present test material. Consider, for instance, how response and presentation modes will affect the child's performance. Do you present items giving both visual and auditory cues? Are you allowing an equal representation of *say, write,* and *do* responses, or for some reason do you find it necessary to concentrate more attention on one of these response categories than another? Have you considered factors such as motivation?

An informal inventory does not bind you to any one form of presentation. You can, if you wish, specify individual testing conditions for each child you test. It is important, however, that you are aware of the way you handle your presentation and that you try to be as consistent as possible.

Deciding on direct or indirect observation techniques. Keep in mind that performance on an inventory is not always directly observed. Some inventories present a set of questions where performance can either be *directly observed* or *reported* by a parent, teacher, or other interested party who should be familiar with the way the child performs. Indirect observation results often provide valuable new information beyond that which you can directly observe in a short time. To administer a language inventory, for instance, the teacher of a developmentally delayed or very young child might devise a set of questions for the parent to answer concerning names of family members and acquaintances the child often uses, favorite toys or preferred foods, TV shows, characters, commercials he talks about, verbs he seems to understand, articles of clothing or household objects he understands, and examples of directions he typically follows as opposed to a list he consistently seems to have trouble with.

Establishing scoring criteria. Before administering your inventory, it is critical that you decide exactly what types of information you hope to glean from testing and exactly how you will define your results.

Many inventory items may be set up to require a simple yes/no response. Consider skills listed in a self-help skill sequence (e.g., dressing: *Does he button his coat, zip, snap,* etc., would require a simple yes/no response). Other items, such as spelling a list of words, reading words aloud from a passage, writing digits in answer to math problems may call for correct/error tallies, with results summarized in terms such as percentages.

Still other types of inventories require that you present an item repeatedly and find out how often the child responds correctly (e.g., he answered his name correctly on 3 of 4 responses). Check your inventory carefully to make certain you know exactly how you plan to score each item before you administer it.

The more precisely you define performance expectations—that is, what constitutes correct and error, the more easily you can evaluate the behavior you observe. Do you want to break out and analyse specific kinds of performance errors, as in reading, where substitutions, omissions, insertions, self-corrections, and mispronunciations may all be looked at separately? In math, you might look for problems such as: (1) inadequate knowledge of basic facts; (2) use of incorrect operation (although the answers are correct); and, (3) ineffective strategy (the child selects the correct operation and provides adequate facts, but is inaccurate because he applies steps out of sequence, skips steps, or applies some tactic which does not result in the correct answer). Recognizing that such error patterns occur will help you establish future programming directions.

As you review your inventory with an eye to scoring, make sure that you have presented enough of each kind of item, so that error patterns, if there are any, have a chance to manifest themselves. And again, decide how you want to summarize information about general response patterns (e.g., defining independent, instructional, and frustration levels of performance).

Planning the scoring mechanics. Is this a paper-and-pencil inventory, for which you need to set up a response form? If not, how do you plan to observe and score the child's performance? Will you sit beside him and score on a separate follow-along sheet? Can you tape-record reading performance for later scoring? Plan your scoring mechanics in terms of what's easiest for you and least intrusive to the child.

Making related observations. Remember, as you administer the inventory, because it is an informal situation, you have a fine chance to observe not only the specified performance, but also any other relevant strategies the child employs, or extraneous behaviors he engages in. For example, use of fingers when doing basic facts or working from left to right in solving two-digit addition problems could be important observations for later programming.

Note any increase in behavior problems (e.g., stuttering or tantrumming) that seem related to the testing situation, and record any specifics of the conditions surrounding the problem behaviors you notice.

Inventory Examples

Since the inventory is an informal testing procedure, its format, administration, and scoring can vary widely. The examples here show the way three different teachers have gone about making up inventories: the first, offering the administrator great flexibility; the second, a more structured approach; and the third, focusing on guidelines for interpretation of pupil response.

An inventory with unstructured conditions. The pre-math inventory this teacher created allows for great flexibility on the part of the administrator. Items are offered with some general guidelines, leaving whoever administers the inventory to adapt the content to the particular child and situation. The teacher bases his assessment upon portions of the pre-math sequence in *The Developmental Resource* (Cohen and Gross 1979, pp. 129-131). The total inventory includes the following sections: Classification Concepts, Ordering Concepts, Cardination Concepts (including Estimating Number of Objects in a Group and Rational Counting), Rote Counting Skills, Numeral Identification Skills, One-to-One Correspondence, Comparison of Set Size, Conservation Concepts, Relevant Number Transformation Concepts (addition and subtraction), Time, Money, Use of Calendar/Ruler/Thermometer, and Early Concepts of Fractions. Only sections involving Estimating Number of Objects in a Group, Rote Counting, Numeral Identification and Rational Counting are included here. (See Figure 3:1.)

FIGURE 3:1. *Premath Inventory: Unstructured Conditions.*

General Directions for Premath Inventory

1. For each of the skills listed, there are a variety of ways you may go about establishing the level at which a child is functioning. The way an item is administered can heavily influence the results obtained. If you are concerned with obtaining maximum information for future programming, therefore, it becomes critical to note not only the child's success or failure to perform, but also under what conditions the item is being administered, making performance possible.

2. Some of the items appearing in this inventory will immediately suggest conditions under which they might be implemented; and, depending on the child you are examining, further clarification may not be needed.

3. Space has been provided beneath each item for you to make any notes where you feel clarification regarding your presentation is necessary or helpful. As you try to determine what information you might want to consider as you present certain items, the following review may be helpful:

   "Who" variables. Will the person who presents the items be a factor; is the child in any way actively involved in the presentation; that is, will he have any control over getting or presenting materials to himself?

   "What" variables. If some type of stimulus material is to be presented, what dimensions of this material might be critical (e.g., its size, shape, texture, color, auditory properties, or reinforcement value)?

   "Where" variables. Might the place an item is presented be a factor (e.g., crowded classroom versus a special testing room or screened area)?

   "When" variables. Might time affect the child's performance? For example, is a certain time of day preferable; might the child perform differently if the item were incorporated as a part of his daily schedule (e.g., counting something he needs to use now, versus a group of abstract objects)?

   "How" variables. How is the item to be presented? Are instructions to be given; how will they be presented—with gestures, words, pictures? What will be said if directions are verbalized? Will there be an accompanying demonstration? Is feedback given?

4. If you decide to compare and contrast performance under two or more conditions, given the same item, note those variables you intend to try.

   Example: Counts 3 objects. (Circle Yes or No.)

a.	Concrete objects (e.g., poker chips)	Yes	No
b.	Pictured sets of airplanes	Yes	No
c.	Pictured dot array	Yes	No

5. The space beneath an item can also be used to make any comments about related behaviors you observe as the child deals with the item (e.g., error patterns, manner of approaching problems, and/or behaviors accompanying his response).

6. Under each major heading, a short list of some possible considerations to include in the space provided has been given.

Subitizing

This skill involves arriving at an estimation of number without overt counting of groups of objects. To begin, items presented should be homogeneous, identical in number, shape, size, and identity. You may later want to try contrasting performance on homogeneous versus heterogeneous groups of objects. Issues to consider include how objects are presented and specific types of objects used (e.g., concrete, abstract, and type of display format used).

1.	Estimates verbally the numerosity of set sizes of 1 to 2 objects accurately.	Yes	No
2.	Estimates verbally the numerosity of set sizes of 1 to 3 objects accurately.	Yes	No
3.	Estimates accurately the numerosity of set sizes of 1 to 4 objects.	Yes	No
4.	Estimates accurately the numerosity of set sizes of 1 to 5 objects.	Yes	No
5.	Estimates accurately the numerosity for sets as large as 6 elements.	Yes	No

Rote Counting Skills

(Hint: Start at end of sequence, item number 5, then locate the place in the sequence which best describes how far the child was able to count before problems occurred.)

1.	Says numbers by rote, generating sequence 1 to 5.	Yes	No
2.	Says numbers by rote, generating sequence 1 to 10.	Yes	No

3. Says numbers by rote, generating sequence 1 to 15. Yes No

4. Says numbers by rote, generating sequence 1 to 20. Yes No

5. Says numbers by rote, generating sequence beyond
   20 (indicating how far the child counts correctly). Yes No

Numeral Identification Skills

Include how a cue is given, how many alternative numerals were presented for the child to consider, and which numerals he had problems with.

1. Matches numerals 1 to 5. Yes No

2. Matches numerals 6 to 10. Yes No

3. Marks the correct numeral, 1 to 5. Yes No

4. Marks numerals, 6 to 10. Yes No

5. Names numerals, 1 to 5. Yes No

6. Names numerals, 6 to 10. Yes No

7. Writes numerals correctly, 1 to 5. Yes No

8. Writes numerals correctly, 6 to 10. Yes No

Seriation

This skill involves placing numerals in correct order. (Include here how presented, what materials are used.)

1. Seriates numerals 1 to 5. Yes No

2. Seriates numerals 1 to 10. Yes No

Rational Counting

Counts fixed, ordered sets. The child cannot move the objects as he counts them. If they are concrete objects, they are presented in such a way as not to be moved. Issues include homogeneous versus heterogeneous sets, pictured objects versus concrete objects, type of pictured object chosen (presentation of real objects versus abstractions such as a dot array); worksheet or flashcard presentation.

Counts

1 . . . . .object	Yes	No
2 . . . . .objects	Yes	No
3 . . . . .objects	Yes	No
4 . . . . .objects	Yes	No
5 . . . . .objects	Yes	No
6 through 20 objects	Yes	No

Counts movable objects. Select objects the child can manipulate or move as he counts. How are objects presented; what objects are used; are objects homogeneous or heterogeneous; how does the child deal with the task—are there any extra behaviors worthy of mention?

Counts

1 . . . . .object	Yes	No
2 . . . . .objects	Yes	No
3 . . . . .objects	Yes	No
4 . . . . .objects	Yes	No
5 . . . . .objects	Yes	No
6 through 20 objects	Yes	No

An inventory with structured conditions. This receptive language inventory (Figure 3:2), based upon Sensorimotor/Early Cognitive and Language Development sequences in *The Developmental Resource* (Cohen and Gross 1979), is a highly structured one for which the examiner is given precise directions concerning test administration.

FIGURE 3.2. *Receptive Language Inventory: Structured Conditions.*

---

1. Turns head at least two out of three times, within 30 seconds, when "Hi, [child's name]," is spoken from approximately three feet, at a 90-degree angle from the child. Test for both ears.

Left ear—turns	Yes	No
Right ear—turns	Yes	No

2. Turns within 30 seconds when various sounds are presented at approximately three feet, 180-degree angle from the child. Randomize presentations.

Soft rubber squeeze toy

Left ear	Yes	No
Right ear	Yes	No

Paper crumpled

Left ear	Yes	No
Right ear	Yes	No

Rattle sounded

Left ear	Yes	No
Right ear	Yes	No

Two blocks knocked together

Left ear	Yes	No
Right ear	Yes	No

3. Gives toy to examiner on request when request is accompanied by gesture. A book, teddy bear, and a toy truck are placed before the child. Let the child touch the objects, then ask for whatever toy the child puts hand on by saying, "Give me the (book), (child's name)," and holding out your hand.

Yes   No

4. Responds appropriately to three different direction commands, given without accompanying directions. Toy bear is placed at opposite end of a table out of child's reach. Command is given but *not* accompanied by gestures.

Get the toy bear	Yes	No
Put it on the paper	Yes	No
Give it to me	Yes	No

5. Points to correct part of body when asked, "Show me your (nose)."

Mouth	Yes	No
Nose	Yes	No
Hair	Yes	No
Ears	Yes	No
Eyes	Yes	No

An inventory with guidelines for interpreting pupil response. The previous two examples allow considerable flexibility when it comes to summarizing and interpreting results for the purpose of setting up individualized objectives. The following inventory (Figure 3:3), based upon the Bucks County 1185 Common Words list, offers specific suggestions for interpretation. These suggestions can later be incorporated in the establishment of objectives for the reading program.

General Directions for Administration

1. Give the pupil a copy of the word list and retain one for yourself. Mark pupil's responses on your follow-along copy as student reads aloud from his.
2. For additional information about errors and the opportunity to later determine common error patterns, indicate incorrect responses by writing the child's response next to the word read incorrectly (e.g., the word to be read is *in* and the child pronounces *on*; write *on* beside the word *in*).
3. Estimate the child's reading level and begin one level below the estimated level.
4. Have the child attempt each consecutive list until he misses 10 percent of the words on any one list; the *frustration level* in word recognition is defined as 90 percent or less average correct.
5. Go back and determine the following:
   Instructional level for word recognition (95 percent correct)
   Independent level (99 percent words read correctly)
6. Compare the pupil's level of word recognition on basic vocabulary words with his reading comprehension skills. If comprehension level is lower than vocabulary level, start instruction on the level of comprehension skills.

Preprimer Word List

a	and	ball	big	blue	can
come	down	father	for	get	go
have	here	house	I	in	is
it	jump	little	look	make	me
mother	my	not	oh	play	red
ride	said	see	the	to	up

Primer Word List

all	am	are	at	away	birthday
boat	box	boy	but	came	did
do	dog	doll	duck	eat	farm
find	fun	funny	girl	good	has
he	help	home	kitten	know	laugh
like	Mr.	no	now	on	one

First Reader Word List

about	after	again	animal	apple	around
as	ask	baby	back	book	barn
be	bear	began	black	brook	brown
by	call	cat	children	cold	color
could	cow	day	door	egg	far
fast	feet	first	fish	fly	four

USING CRITERION-REFERENCED TESTS TO ASSESS PERFORMANCE

The major difference between the criterion-referenced test and the inventory is that behavioral objectives for the criterion-referenced measure are formulated *before* evaluation takes place; that is, test items are based upon predeteremined objectives. In the case of the inventory, on the other hand, performance serves as a basis for formulating behavioral objectives, with the objectives specified only after performance and evaluation takes place.

While scores on criterion-referenced tests may be stated in terms of percentages similar to those achieved on norm-referenced tests, the interpretation of these scores is quite different. For example, a score of 70 percent on a criterion-referenced test indicates that the student needs instruction on the remaining 30 percent. A score of 70 percent on a norm-referenced test indicates that 30 percent of the tested population answered more items correctly than the student, without reference to prespecified objectives.

Criterion-referenced tests have the advantage of organizing evaluation results in terms of specific objectives and criteria. Rather than speculating about what skills to teach a fourth grader performing at a 1.5 grade level in reading, the teacher can simply transfer behaviorally stated items from the test record form to daily plan forms. This advantage is somewhat offset by the fact that clearly stated, clearly organized objectives and criteria take some time to formulate. These objectives and their criteria are the crucial groundwork for the creation of any criterion-referenced test. Lack of sufficient time and planning for test development can result in a hastily constructed device that does not adequately represent the educational performance of the student, or the educational requirements of the teacher.

Guidelines for Creating Criterion-Referenced Tests

While criterion-referenced tests vary considerably in terms of the subject areas with which they concern themselves, the numbers of objectives they examine, and the format this examination takes, certain issues are common to all.

Defining objectives. Implementation of the criterion-referenced test usually indicates that you are setting up an environment in which objectives are specified, and learning is monitored to assure that those objectives are met.

Since the objective is the focal point of every criterion-referenced test item, it stands to reason that having clear objectives in mind will be very important to you. It helps if the objectives you are using are organized in some way (e.g., according to a master curriculum plan, behavioral or developmental sequence). The better organized your objectives are, the better able you will be to use them as an integral part of cohesive program planning.

Defining criteria. Make certain you are able to state clearly the criterion levels you expect students to meet. Most criterion-referenced tests are informally constructed by teachers with objectives and criteria related to their particular situations and curriculum. You can, however, find some assistance in setting criteria by looking to state, district, textbook publishers, or authorities in a particular curriculum area for guidelines in establishing desired levels of performance.

Remember that the way in which you state a criterion will determine the way in which you perceive the child's performance on the test. If the child is to identify the hour on a clock face, with 100 percent accuracy as criterion, and you have given him only one item in which to show understanding, he will score 100 percent simply by answering one item correctly. (See Chapter 4 for more information on establishing criterion levels.)

Administering the test. Most often, a criterion test will be administered twice: before instruction begins, to determine where the child performs in relation to established criteria, and again, following instruction, to determine whether the child has achieved mastery. The child continues to receive instruction and retesting until criteria are completely met.

Relating test items directly to objectives. Test items must be directly related to objectives you have established. Ask yourself what kinds of items would require performance demonstrating that the child, in fact, does understand the material. Can you say, "I would expect a child who completed my instruction successfully to be able to handle this type of item successfully"?

Selecting test items. Remember to have at least one test item keyed to each of your objectives, and remember, conversely, that each item on the test must represent a specified instructional objective. Items on an arithmetic inventory, for instance, should relate directly to such performance categories as basic addition and subtraction facts, sums 0-20, time-telling, and coin recognition, rather than to a specified grade level of performance. As was mentioned earlier, relating items directly to objectives guarantees that when results are evaluated, the teacher is able to tell precisely what skill category to work on, and even what specific kinds of tasks need attention.

Examining test length. It may be to your advantage for ease of administration during a busy day in the classroom to have a few shorter tests covering similar skill groupings than one long test covering a wide range of skills on an entire program sequence for an area. The shorter tests also take into account the fatigue factor which can well affect child performance.

Grouping items. Determine if objectives are related to each other in a cohesive way. Try to group those objectives which are related together along with each of the items which have been keyed to them. Attending to such organization ahead of time makes later interpretation much easier.

Consider the effects of presentation and response modes. Ask how much variety in presentation and response modes is required. If you are requiring everything from manual demonstration to oral or written responses on the same test, you may find some problems in administering this evaluation easily and efficiently.

Try placing objectives requiring similar instructions and modes of responding together on one test for ease of presentation.

One of the major advantages of an informal test is its flexibility. It allows you to alter certain ways the test is administered, making it possible to examine differences in child performance, as well as allowing the child his best chance to show you what he is capable of doing. As a result of the work involved, you will want to design the test with a broad range of students in mind; but as you administer it to an individual child, keep in mind such factors as ways the child prefers to respond (e.g., oral versus written), what way items and/or instructions might be most effectively given, what his native language is, and if there is any type of handicap which should be taken into account. You might want to compare and contrast performance as you allow for some of these factors in your presentation.

Criterion-Referenced Test Example

The criterion-referenced test in Figure 3:4 is based upon a small portion of the premath sequence in *The Developmental Resource* (Cohen and Gross 1979, pp. 129-131). Note that objectives have been clearly specified *beforehand*, with test items developed and keyed to each of the objectives. Criteria for mastery have been predetermined, and mastery is only considered to have occurred if the child meets those criteria. Resulting test performance tells the teacher which skills have been mastered and which have not.

The criterion test does not permit the flexibility possible with administration of the inventory: conditions are specified carefully for each objective and the test items based upon that objective.

FIGURE 3.4a. *Criterion Referenced Premath Test Objectives.*

Part I

Rote Counting Skills

Says numbers rotely, generating number sequence to 20, when asked to count from 1, with 100 percent accuracy.

Numeral Identification

1. Matches numerals when given a model on a 3x5 card and three randomly selected alternative 3x5 cards, only one of which is a correct match, with 100 percent accuracy. The following numerals are to be used and presented in random order:

1	6
2	7
3	8
4	9
5	10

2. After hearing a number read and asked to show one of three randomly selected alternatives on the sheet before him, identifies, by marking the correct numeral with 100 percent accuracy. Present numerals 1 through 10 in random order.
3. When presented with cards showing each numeral, names numeral with 100 percent accuracy. Present numerals 1 through 10 in random order.
4. Upon hearing a number as it is dictated to him, writes numeral with 100 percent accuracy. Present numerals 1 through 10 in random order.

Seriation

Given ten 3x5 cards in random order, each with a numeral from 1 through 10 printed on it, places cards in their correct order with 100 percent accuracy.

Part II

Rational Counting

Counts fixed, ordered sets.

1. Given ten different sets of plastic animals (heterogeneous groupings), each mounted on a cardboard base, counts aloud with 100 percent accuracy. Sets are to represent the following numbers presented in random order:

1	6
2	7
3	8
4	9
5	10

2. Given ten different sets of poker chips (colors homogeneous), each mounted on a cardboard base, counts aloud with 100 percent accuracy. Sets are to represent numerals 1 through ten, presented in random order.

3. Given ten different heterogeneous sets of pictures arranged on a worksheet, counts aloud with 100 percent accuracy. Sets are to represent numerals 1 through 10, presented in random order.

4. Given ten different dot arrays arranged on a worksheet, counts aloud with 100 percent accuracy. Sets are to represent numerals 1 through 10, presented in random order.

FIGURE 3:4b. *Criterion Referenced Premath Test.*

Part I

Rote Counting Skills

Say "Now we are going to count. Let's begin . . . 1 [pause and let child go on] . . . " Say "stop" when child reaches 20. Note errors and omissions in sequence.

Numeral Identification

1. Match numerals

   Material: Twenty 3x5 cards; each numeral is clearly printed on two cards. Separate the cards into two stacks, each containing cards with the numerals 1 through 10.
   Setting up: Pull one card out of the "model" stack. As you "shuffle" the other stack, draw the matching card and two distractors from your pile. Place the matching card and the two distractors in front of the child.
   Teacher says and/or does: Hold up the model card. "Can you put this card on top of the one that looks just like it?" Hand the card to the child. Repeat this as you present ten randomly selected numbers; place a "c" by those the child identifies correctly.

1	6
2	7
3	8
4	9
5	10

2. Mark numerals

   Materials: A master sheet with the numbers 1 through 10, order randomly selected, is needed for dictation. The master sheet is keyed to a worksheet which presents ten framed boxes. Each box contains three numerals, one a correct match with the number on the master, and two randomly selected distractor numerals.
   Setting up: The child needs a worksheet and a marking tool (pencil, marking pen, crayon). The teacher needs the master sheet.
   Teacher says and/or does: Put your finger on the first box. "Mark the numeral three." Repeat for each numeral. Place a "c" on the master beside correct responses.

3. Say numbers

   Materials: Ten 3x5 cards each with one of the numerals 1 through 10 printed on it.
   Setting up: Shuffle cards. Place randomly ordered stack on table before you.
   Teacher says and/or does: "I'm going to show you some cards. You tell me what numbers they say." Hold up the first card. "OK, what's this number? . . . and this? . . ."

Place a "c" by those the child answers correctly.)

4. Write numerals

   Materials: A master sheet with the numbers 1 through 10, order randomly determined, is needed for dictation.

   Setting up: A piece of unlined 8½x11" paper with ten boxes arranged vertically for answers should be given to the child along with a writing implement.

   Teacher says and/or does: "I'm going to read some numbers to you, and as I read, I want you to write them in the boxes on your paper. Show me the box on the top of your paper. We'll start with this box. Write *1*. Now find the next box. OK—write *5*." Place a "c" by those the child writes correctly.

Seriation

Materials: Ten 3x5 cards each with a numeral from 1 to 10 printed on it.

Setting up: Place the cards in random order in a stack face down in front of the child.

Teacher says and/or does: "On each card in front of you there is a number. These cards are all mixed up so the numbers are all out of order, something like 3, 8, 5, 7 instead of 1, 2, 3, 4. When I say 'Begin,' I want you to turn the cards over and see if you can fix them up the right way. See how many you can put back in the right order. OK—begin." Indicate how far the child was able to go before a problem in sequencing appeared.

Part II

Rational Counting

Counts fixed, ordered sets

1. Say numbers when counting concrete representational objects (e.g., plastic animals).

   Materials: Ten different heterogeneous sets of plastic animals each mounted on a cardboard base and representing the numbers 1 through 10.

   Setting up: Place the sets of animals in random order and seat the child at a desk or table.

   Teacher says and/or does: "I'm going to show you some animals. Each time I put a group of animals on the table, I want you to tell me how many you see. Here is the first one. Count how many animals you see." Each time the child finishes counting, place a new set before him and ask him to count how many there are. Put a "c" next to those sets he counts correctly.

2. Say numbers when counting concrete counters (poker chips).

   Materials: Ten different sets of plastic poker chips, each mounted on a cardboard base and representing numbers 1 through 10; colors for each set are homogeneous.

   Setting up: Place the sets of poker chips in random order and seat the child at a desk or table.

   Teacher says and/or does: "I'm going to show you some chips. Each time I put a group of these chips on the table, I want you to tell me how many you see. Here is the first one. Count how many you see." Present each set. Put a "c" next to those he counts correctly.

3. Say numbers when counting abstract representational objects (pictures of objects).

   Materials: Ten different sets of heterogeneous pictures representing numbers 1 through

10, randomly ordered and arranged on a worksheet. Frames are placed around each set.

Setting up: Place the worksheet in front of the child.

Teacher says and/or does: "This page has a lot of different pictures in each box. Let's start here (point to the first box) and tell me how many you see in this box." Present each set. Place a "c" next to those counted correctly.

4. Say numbers when counting abstract counters (dot arrays).

Materials: Ten different dot arrays, representing numbers 1 through 10, are randomly selected and arranged on a worksheet. Frames are drawn around each set.

Setting up: Place the worksheet in front of the child along with a writing implement.

Teacher says and/or does: "This page has dots in each box. Let's start here (point to the first box) and tell me how many you see in this box." Present each set and place a "c" next to those he counts accurately.

Skill areas to be tested are chosen selectively, omitting those which may be interesting, but for which the teacher is either not able or not inclined to set criteria. For instance, subitizing (estimating) skills which were included in the earlier premath inventory, are excluded in this criterion-referenced test. The teacher may have felt that a skill, such as the child's ability to estimate number without counting, is interesting, but not something on which he required a certain level of performance. For the same reason, rote counting is required only up to the number 20. While some children in this teacher's classroom do count past 20, the skill is not required for this group and, therefore, no criteria are given.

The teacher who has created this test has established a criterion level of 100 percent for each item. It is important to realize that a number of factors are involved in setting a criterion level. In another situation, 100 percent might not be applicable, and, in fact, a variety of different criterion levels might be used throughout a listing of objectives. Each objective needs to be examined individually to determine appropriate levels of performance. Chapter 4, describing educational objectives, explains this issue.

USING PROBES TO ASSESS PERFORMANCE

The probe concept of assessment is an outgrowth of teachers' efforts over many years to devise better informal assessment tools, yielding results geared more intensively toward individualized programming. Probes are designed to interlock tightly with individual instructional units; and, in fact, many look almost exactly like worksheets teachers have long been designing to supplement traditional programs. The probe differs from other informal assessment devices in the depth and detail of its focus upon performance within circumscribed skill areas, as well as in the number of performance opportunities the child is given during assessment.

Probe Characteristics

Probes are unique in their conscious recognition and maximization of elements teachers have often intuitively included in good informal assessment procedures. While individual probes may take a variety of forms, three characteristics especially differentiate them from the informal tools already described.

Each probe examines performance in relation to a single skill area. While criterion-referenced tests and inventories may sample performance in regard to several different areas and across a wide range of highly diversified skills, the probe always focuses on a single skill area and on one particular topic, issue or task. The assessment tools we have discussed up to this point ask numerous kinds of performance questions, but the probe forces us to ask just one question at a time.

For example, a third-grade math inventory might survey a child's understanding across a number of different arithmetic skill areas (e.g., basic computation, telling time, concepts of money and measurement). For each area assessed, four or five items could be considered sufficient to represent the child's skills; and the items themselves might be extremely dissimilar. Computation, for instance, might require addition, subtraction, multiplication, and division. Contrast this type of assessment approach with the probe shown in Figure 3:5. Not only does the probe focus on a much narrower segment of performance (in this case, multiplication problems of the "single-digit times double-digit"), but it gives the child opportunity to demonstrate the called-for skill over a much larger number of items. While the math inventory we have described might contain one or two problems of this particular type, the probe shown here contains 80 such problems.

Each probe examines performance in terms of a single specified behavior. Other informal assessment measures often encourage a variety of response types (such as circling answers, writing answers, filling in the correct letter alternative, or providing a verbal response). Each probe calls upon only one particular type of behavior. In the example cited in Figure 3:5, for instance, the probe is constructed to require *write-digit responses* throughout the entire sample. Focusing on one behavior alone is based on a rationale that scoring and evaluation will be easier, and that changes in performance will be more likely related to skill difficulty, rather than to differences in the ways the child is asked to respond.

Each probe examines performance over a number of days. In addition to allowing the child numerous opportunities to encounter a single problem type within a given probe, each probe (or a variation of it) is intended to be given repeatedly over several days. The assessment instruments considered thus far are used most often as "one-shot" measures. They examine performance *one time* to determine the child's performance capabilities. A probe, on the other hand, asks not only what the child presently knows but also what the child *could learn* about this particular material if he were given the opportunity to interact with it on several different occasions.

Administering the same or comparable material several days in a row results in a completely different type of performance picture: what we call a *performance profile*. Such a profile provides a very clear portrait of the child's abilities in relation to the pinpointed skill, and, at the same time, gives us a glimpse of one important aspect of the learning process itself. That is, it allows us to see the effect of giving the child repeated opportunities to see and practice the particular behavior a specified objective requires.

Here are a few of the performance profiles which commonly emerge when probes are administered:

1. Child A does rather poorly on the first day with few correct answers and several errors; over repeated presentations of similar probes in a five-day period, he shows a steady reversal in this pattern, ending with many more correct answers than errors.
2. Child B does pretty much the same from one day to the next.
3. Child C does well on the first day and then successively worse with each repeated presentation.
4. Child D shows erratic performance, each day radically different from the one preceding it—one day a high number correct; next day, an extremely low number correct, etc.

FIGURE 3:5. *Multiplication Probe: Single Digit x Single Digit.*

Student: _____

Age: _____

School: _____

Grade: _____

Date: _____

One-Minute Sample

4	6	2	1	7	8	3	7	8	5
x9	x6	x3	x0	x8	x5	x4	x2	x1	x7

7	3	3	7	4	9	5	3	2	4
x8	x0	x3	x6	x3	x0	x0	x6	x7	x8

8	5	6	6	2	7	6	3	3	0
x7	x9	x5	x3	x0	x4	x4	x9	x8	x0

8	5	4	6	4	0	1	7	8	7
x1	x2	x7	x8	x6	x0	x4	x5	x9	x3

6	5	3	3	3	4	9	8	2	1
x1	x8	x7	x5	x2	x5	x9	x6	x3	x9

6	1	2	6	9	2	8	2	4	3
x9	x3	x9	x7	x8	x6	x3	x5	x2	x9

5	3	0	8	6	8	5	7	6	9
x4	x9	x8	x8	x3	x3	x6	x9	x7	x7

9	1	7	3	2	8	3	9	3	7
x9	x2	x8	x4	x9	x4	x6	x5	x9	x7

In addition to showing how the youngster presently stands in relation to the objective you have established, these results clearly suggest possible programming strategies. Child A appears to benefit from repeated opportunities to practice. Child B is extremely consistent, with practice alone not a factor in helping him improve; he will require further instruction if improvement is desired. Child C may respond to the novelty of situations, making drill and practice alone a deadly situation; introducing novel ways to go about practicing a skill when such practice is needed may be necessary. (A look at some possible incentives and interest changes might also be helpful here.) And Child D may require a further examination of just exactly what is happening in his environment; the child's performance is "out of control." A complete examination of contingencies and reinforcers available would be an excellent place to start.

The ability to identify a "learning profile" in regard to a specific skill is obviously an advantage. Probes entail other advantages as well. Because they can so easily be designed within the well-known worksheet format, teachers often feel more comfortable with them than with other, less familiar, assessment instruments. Yet another advantage is that probes can serve a variety of functions in addition to their primary purpose, which is, of course, initial assessment prior to instruction.

- Probes can be used as a part of instruction, acting as a kind of worksheet to provide additional skill practice.
- The probe can be used at intervals throughout instruction to provide a periodic check on how the child is doing in relation to both initial performance and desired terminal skills.
- A probe identical or similar to that used originally for assessing performance can be administered at the completion of instruction to determine whether or not initial performance criteria have been met.

Guidelines for Designing Probes

Teachers involved in implementing the collection of measurement procedures that have come to be known as Precision Teaching—and, most recently, *Exceptional Teaching* (White and Haring 1980)—were responsible for creating the notion of probes as an assessment tool. They were eager to develop an assessment strategy that would allow each child to respond in a way most revealing of his own individual performance style, as well as to allow children to give their best performance, free from the barriers that many other assessment devices, however unintentionally, place in their way. You may find the guidelines they established useful to you as you begin designing your own probes. Remember, however, that they are *only* guidelines. While the three features elaborated initially (i.e., examining performance in relation to [1] a single objective, in terms of [2] a single behavior, and [3] over a number of days) are absolutely essential to the nature and functioning of a probe, the issues below are merely suggestions which you may choose to follow or disregard as you wish.

Selecting item type. The problems selected for a probe should be representative of the skill you wish to assess. This is a suggestion with a double hook in it: (1) be sure to include *only* problems which sample the skill you are interested in (e.g., a multiplication probe examining single-digit times single-digit facts would not include any items containing double-digits in either the multiplier or multiplicand positions; and (2) try to include items that will represent the *entire range* of skills you are assessing.

Consider, for example, a probe covering the multiplication facts, "times zero" through "times nine." In order to include a fair and representative sample of all problems in that range, without giving undue emphasis to any part of the spectrum (e.g., more "times eight" than "times three"), you will need to employ some strategy that ensures an equal number of all item types, distributed evenly throughout the probe. The use of a table of random numbers is one way to make sure that item selection is not weighted toward one particular item type over another.

Item selection also depends upon whether the probe you are creating is global or finely sliced in terms of content.

The kind of probe just described (e.g., all multiplication facts, "times zero" through "times nine") is fairly global in nature. A more finely sliced probe might contain only those facts from "times zero" through "times five"; or even finer yet, facts within a single times table, such as "times one."

Any probe can run a broad spectrum. The finer you slice, the more directly you can identify specific problems the child may have. In general, a good strategy to employ is to start with the global probes, going back to the finer slices as a need is indicated.

Deciding on number of items. Traditional assessment devices limit the number of opportunities the child has to demonstrate his skills on a particular type of problem. For instance, an entire test may contain no more than three or four of a single item type. Sometimes only one item is given to represent an entire class of responses. Probe originators felt that such testing strategies "locked in" the child's performance. That is, no matter how well the child responded to the items given, he could do no better than the limits imposed by the test itself. Testing often resulted in a picture of performance which said more about the teacher's or examiner's expectations than it did about the child's own learning capabilities and style.

For this reason, probes were initially designed to "unlock" performance in a special way, by always giving the child "unlimited" opportunities to respond. Of course, no assessment tool can really present an *unlimited* number of items, but probe creators saw that they could simulate such a condition by making sure that a child always encountered more items than he could possibly complete in the amount of time allowed.

For example, the teacher might construct a multiplication probe to contain 80 or more similar problems. The child is allowed a one-minute timing in which to work the problems as well as he is able. The teacher emphasizes that everyone works differently and stresses that *no* child is expected to do *all* the problems. Each simply completes as many as he can. Since it is highly unlikely that a youngster who typically does about 40 such problems in one minute would complete all 80 items, the child would essentially be confronting an "unlimited" number.

Probes traditionally have limited performance time, then, but not the number of problems or items a child could conceivably solve in the specified time. Probe originators found that by "unlocking" a child's performance in this way, they were able to obtain a much more detailed picture of each child's individual performance style. Given a virtually unlimited chance to demonstrate his skills, how does the child respond? Some children, confronted by numerous opportunities, work extremely quickly, doing as many problems as they can. Having a large number of items in front of them seems to spur them on. Other youngsters may demonstrate a very slow and plodding performance. Although speed may not be considered the most critical performance component, it is certainly a factor in almost all learning or working situations.

Accuracy also may be an issue when a child is given a large number of items to complete. The child who, when confronted with only three items on a traditional assessment tool, solves all three accurately, may continue to work accurately or may demonstrate a much different pattern when he is challenged with 80 items. If the child is told to skip over those problems he feels unsure about, an additional picture is gained of those types of items on which particular problems occur. By giving the child more opportunities to perform, the teacher is able to sort out the accurate plodders, the accurate speeders and the slap-dashers in a way that a very limited number of items never allows. At the same time a teacher is often able to identify areas of instructional difficulty that may not have come to his attention on a much smaller sample of items.

Finally, by looking at the child's performance over several days, the teacher is able to see how that performance changes if the child is given the opportunity to do more and more items each time. This is the kind of learning profile discussed at the beginning of the probe section. It carries with it a capability for demonstrating performance change not possible when response opportunities are limited. A learning profile lets us know much more about the child than merely his response to a select set of test items. It clues us in also to potential learning characteristics that may accompany a youngster throughout his educational career. Characteristics such as speed, accuracy, proficiency, and endurance are important to the child in the competitive world of the classroom and beyond. Even more important, perhaps, is the idea of encouraging teacher and pupil alike to view learning as an ongoing *process* which far transcends any isolated body of information that may be covered at a single point in time.

If you do decide to set up probes in this way, giving the child "unlimited" opportunities, you will need to determine just how many chances to respond are more than enough. One way to do this is to see how many items you yourself can complete in the specified amount of time. Another way is to make a guess at a child's probable upper limit, or to use the established limit of the top child in your class. If the best multiplier in your room does problems accurately at a rate of 60 per minute, always put more than 60 on a probe, making it possible for even the best to be challenged.

Item arrangement. Arranging problems in an organized fashion on the probe sheet makes correcting and scoring easier. For example, laying problems out in units of 10 across the page makes it much easier to correct problems and to count up total correct and error responses. If you are counting a movement cycle such as *writes digits*, you may want to arrange problems so that both problem number and total digit number are consistent across each row. For instance, on a math-probe page containing problems combining mixed one-digit and double-digit answers, you might try arranging eight problems per row, randomized with some one-digit and some double-digit responses, so that the total digits counted per row always equaled 10.

By including a sufficient number of randomized items on a probe sheet, you make it possible to reuse the sheet over a number of days with little chance that children will "catch on" to item arrangement. By using laminated probes and crayons or felt pens, or by folding or cutting off problems that have already been completed, you can let a child begin working where he left off the day before, saving yourself probe materials and the work of making up several versions of a single probe.

Time limit. If you decide to limit the amount of time the child is allowed to respond, leaving the number of responses free to vary, you'll need to decide on probe time limits. Decide first how much time you feel you can afford to devote to administering probes each day. Try to select a period that will allow maximum information for the minimum amount of time spent. Teachers frequently choose one-minute, five-minute, or ten-minute sampling periods. Even though one minute may seem an incredibly short amount of time to spend assessing a skill, for many instructional areas, a one-minute sample is not only adequate, it is often the best. Consider skills such as word recognition, basic computation, and spelling. When the child confronts situations requiring such skills outside the classroom, he is usually called upon to respond with speed and accuracy, rather than to engage in a long, grueling performance marathon. For instance, he reads a sign in a store window as the car he is riding in flashes past; he quickly skims the directions on a model kit before setting to work on it; he figures up the total of his allowance and determines how much more he needs for a desired purchase before accosting his father for a little additional cash. The one-minute sample is excellent for assessing those types of skills whose generalization outside the classroom context requires speed and accuracy. Remember, too, that fatigue is often a factor negatively influencing performance on long tests. For both these reasons, it is to your advantage to keep probes as short as possible and yet long enough to gather an adequate sample of the specified performance.

Any attempt to set time limits must also, of course, take into account what the particular items you are presenting involve. For example, discussing how to design an experiment would likely take a youngster much longer than one minute. For some handicapped youngsters such self-help tasks as tying a shoe would not be accomplished easily in brief periods such as one minute, let alone provide opportunities for repeated tries at this task. Certainly, a child writing sentences to answer comprehension questions requires more time than a youngster simply asked to write single-digit answers to basic addition or subtraction facts, regardless of age or skill level. In order to get an adequate sample of a larger behavior unit, such as *writes statements or sentences*, a longer probe time limit is necessary.

Finally, choose a time sample that best suits the requirements of the task, the skills of the child, and that will be easy to repeat over several days without interrupting a busy classroom schedule.

Number of days for assessment. A unique feature of the probe is the stress upon administering it over several days rather than as a one-shot measure, typical of all other assessment devices discussed.

Opinion on the number of days you should try to administer a probe varies greatly within a general range of from five to eleven days, with three days as an absolute minimum. These should be consecutive days or as close to that schedule as possible. Of course, if the child has absolutely no clue about the responses required—seems unable to provide any correct responses on the first day, and unlikely to improve on subsequent days—be sure to avoid practicing bad patterns and the subsequent frustration likely to result. Consider an alternative probe on easier material, or begin teaching at the present level, testing with the probe as you go along.

Presenting the probe. Three factors in the way the probe is presented will affect whether or not the child is able to respond optimally.

The first of these factors involves the instructions you give. Attempt to tailor your instructions so that they contain only that information which the child absolutely requires to perform. When necessary, give a short demonstration to show exactly what you expect. But, be careful to avoid extraneous information that may only confuse the child. Because this is an informal test, it has the advantage of allowing you to introduce a variety of procedures in your testing protocol to establish individual needs. In order to get at this information systematically, you must, however, be consistent in the way you explore such variables or factors and be willing to stick to the protocol you establish in administering your probe from day to day.

A second crucial factor is the probe's *mode of presentation.* Consider whether the way you have chosen to present your probe will facilitate the child's giving his best performance. Try to take into account what you know about the child's preferred modes of presentation (e.g., if a child has a hearing loss, you would not rely on a verbal presentation but give as many visual and gestural cues as possible, including the use of manual English or other sign language forms; if there is a visual handicap, verbal cues will be extremely important). Even though there may not be a sensory handicap, the child may perform differently given different ways of presenting items, so you will want to compare and contrast your results. The information you gather could have important implications for later programming.

A third factor to consider is whether or not you are pacing the child's performance by the way you present items. If so, you may be locking him into a performance level much lower than he is capable of achieving.

> Take, for example, spelling. Do you assess performance by reading the list of spelling words aloud—sometimes stopping to use the word in a sentence, saying it again slowly, and then waiting for all of the children in the group to finish writing before going on to the next word? If this is the case, the assessment you are giving is not really well tailored to test each individual child's abilities and needs. Many children have to sit idly while the slower children write their words.
>
> Although accuracy is traditionally an important measure for spelling assessments, a child also has to be able to spell words *quickly* if he is going to apply them with ease in writing let-

ters, papers, reports, and other written material. In this case, you have learned no more about the child's spelling capabilities than your presentation will allow. Some children could spell two or three times the number of words in the same period, if given a chance. Allowing them the opportunity to do so tells you much more about individual strengths and weaknesses than assessing them on the basis of the ten words that everyone was given to spell.

Teachers have devised several strategies which help to avoid these problems—for example, reading words aloud, or presenting them by means of tape recording at a predetermined pace slightly faster than the child can write. The child selectively picks words and writes them as quickly as he is able. Another approach is to hang a large list of words at the front of the room and ask your student to copy it. Here, spelling becomes a "see-to-write" task, as opposed to the traditional "hear-to-write" assignment it has often been. As his *write letters in words* approaches his skill in writing letters in isolation (i.e., simply copying random letters), he will be working at such a pace that he has time only to glance up and remember what he has seen.

Another example of locking the child's performance by the mode of presentation chosen includes the use of flashcards; your rate, or the child's rate of turning over cards, influences how many opportunities the child has to respond.

There are certain cases in which you want to lock a child's performance . . . where you don't want him to go faster than a predetermined pace or deal with more than a set number of problems. Make sure, however, that when you "lock in" or "pace" a child's performance, you have made a conscious decision to do so. Since locking is so common a way of handling many tests, you may have to reorient your thinking considerably in order to accomplish this. Be patient. Keep asking yourself if you are in any way locking the child's performance by what you say or do—have the experiences offered set the upper or lower limits of performance?

Determining a mode of response. As in presenting any other task, when you present a probe, you must consider whether or not the child is locked into a certain performance profile as a result of the mode of response you are requiring of him. Often, children can work much more efficiently— thus encountering more learning opportunities—*if* they are allowed to use a response mode which is less difficult for them. A child may make numerous errors using one mode (e.g., *say number answers* to arithmetic problems), but answers quickly and accurately on the same material using another mode of response (for instance, *writing answers*). Another child with a handicapping condition may find it burdensome, perhaps all but impossible, to offer a written response, but may be able to respond with comparative ease when given the opportunity to say or point to answers. Be careful to consider any possible implications the response mode you have chosen might have for the performance of the individual child you are assessing.

Identical versus equivalent probe formats. Whether or not you administer the same probe from day to day depends upon the pinpoint content you are examining. If you are calling on simple recall responses, for example, identical probes may work for you, provided you consider ways to avoid memorization of response patterns or sequences. If your probe contains more items than the child can handle in any one session, you may have him start in different places on the sheet you have designed or in the sequence of items (cards, concrete sets of material) you plan to present. The identical probe also works ideally in situations where the question is an open-ended one, requiring that the child provide as much as he can in the time allowed (e.g., creative writing) with each day's goal being that he expand his response somewhat.

In many cases, it will not be possible to use the same probe repeatedly. Generally, in order for you to examine behavior above the recall level, you will need to see that the child has comparable but different items presented from day to day; the format is equivalent, but content changes.

Setting and relevant materials. You don't need to confine your probes to pencil-and-paper activities. They should involve the most relevant materials (e.g., articles of clothing for a dressing program, money for a change making program, food for a cooking program) offered in the setting or situation most relevant to their use to meet criteria for your instruction (a kitchen, bathroom, store setting, etc.). Be certain that you select and make arrangements for all materials and the setting needed as part of the planning of the probe.

Identification information. Don't forget to include a place on the probe where information such as (1) the child's name; (2) teacher; (3) school or special class identification; (4) time; (5) date; (6) summary of results (e.g., number of correct and error responses can be noted). Because the probe is administered over more than one day, and often to several children at a time, such identification information will be important. You may also find it desirable to develop a coding system allowing you to indicate on each probe exactly where the material fits in a filing or other organizational system.

Filing and organizing probes. If you are planning to write a series of probes, try to develop some sort of master plan for coding these probes to a hierarchy of skills, to a scope and sequence chart, or to the behavioral checklist you are using as a basis for your program. Coding and organizing probes is especially important if you plan to pool your efforts with colleagues, creating these assessment devices according to a master plan to which you all agree. Although probes do require an initial time investment, if you are careful to consider reproduction and organization factors at the outset, you will find yourself able to re-use the results of your efforts again and again.

Probe Examples

Since probes are still relatively new in comparison with other assessment devices presented in this chapter, you will find few examples available commercially. One of the commercial probe products that can be obtained is a carefully researched instrument called *Classroom Learning Screening* open space with respect to following revision and deletion—
(Koenig and Kunzelmann 1980), available through Charles Merrill. Since this effort was initiated, numerous school districts and some state departments across the nation have developed their own probe series to be used in the classroom.

Because probes are an informal assessment device, they allow a broad latitude in terms of format and application. By far the most frequently used, especially in Precision and Exceptional Teaching settings, has been the one-minute sample type, giving youngsters a limited amount of time to confront and work with what amounted to an unlimited number of items. Many other options, however, are open to teachers. A few of these options are described in detail in the following examples, showing how teachers working in a wide range of classroom situations have developed and implemented probes. Fundamental to all of these are the three basic probe characteristics outlined earlier:

1. Examination of performance in relation to a single skill area;
2. Examination of performance in terms of a single specified behavior;
3. And, finally, examination of performance over a number of days.

Contrasting two simple probes: Fixed vs. unfixed item number. Figures 3:6 and 3:7 contrast two different ways in which probes might be set up to assess the same skill. The probes are based upon a premath sequence from *The Developmental Resource* (Cohen and Gross 1979), which was used earlier as a source for inventory and criterion-referenced test items. On the first probe (Figure

FIGURE 3:6. *Probe with Fixed Number of Items.*

Name: _____

Probe: _____

Behavior: _____

Criterion Level: _____

Date: _____

Time: _____

Total Correct: _____

Errors: _____

Materials: Strip of paper on which numerals 1 through 10 are printed in random order.

Directions: Ask the child to point to each numeral and say it. "If you don't know one, skip it and go on to the next."

Scoring Sheet: Place a "c" by those the child gets correct.

1	6
2	7
3	8
4	9
5	10

3:6), a fixed number of items—in this case, ten—is offered, with no specified time limit. The teacher simply counts correct and error as the child reads the numerals, and may or may not time the response, depending upon the performance measure chosen. The second probe worksheet attempts to "unlock" performance by giving the child as many chances to respond as possible within a fixed amount of time (one minute). On this sheet, each group of ten numerals has been placed in random order, so that the child will have repeated opportunities to try all ten. Not only is there a chance to name each of the numerals, but, by providing more than the child might possibly do initially, the teacher gives the child room to grow each day, giving the teacher a chance to view other dimensions of performance which may have implications for future programming. There are enough items on this sheet so that the teacher is able to reuse the same probe each day, simply by changing the starting place from one day to the next. If the child finishes the sheet before the specified 1-minute timing has elapsed, he is requested to start over (Figure 3:7).

Probes may be administered in one of two ways. Performance requirements may be established prior to administration, or criteria may be established after learning profiles on probes have been thoroughly examined.

Creating a probe sequence. Probes are used optimally if they are keyed to fit within a carefully planned sequence of instruction. This means designing a sequence of probes for use at multiple levels as youngsters progress from one skill level to the next. Consider the probe sequence designed by Nigel Archer, the teacher of a primary special education class, whose children work at a level above the premath sequence just described. He has created a pencil-and-paper probe set to assess his youngsters' skills on basic addition facts. The probes follow the instructional sequence of which a portion is shown in Figure 3:8. The terminal behavior for the entire sequence has been identified as *writing answers to random addition combinations,* sums zero to 9. As the figure illustrates, the

Name: _____

Age: _____

Date: _____

Admin. by: _____

Time: <u>1 minute</u>

Score: _____

Suggested directions: "This page shows you a lot of numerals, more than we will be able to get through today. When I say 'Begin,' start here [point to upper lefthand corner, first numeral, first row] and tell me as many of these numbers as you can. If you're not sure of a number, just say so and go on to the next one. We're just going to do this for one minute, so don't be surprised when you hear me say 'Stop.'"

5	3	1	8	7	6	2	10	4	9	1	8	6	2	10
3	5	7	9	4	5	1	3	6	2	7	9	8	10	8
2	4	3	5	7	6	1	5	6	8	10	4	1	7	9
3	2	5	6	8	5	7	2	3	9	6	1	4	2	4
5	1	8	9	3	7	6	10	2	4	7	8	1	5	6
9	1	8	9	3	7	6	10	2	4	7	8	1	5	6
9	10	3	1	2	6	3	5	4	8	9	7	10	2	1
3	5	4	7	6	9	8	10	2	4	3	5	9	1	10
7	8	6	3	5	1	2	3	4	7	6	8	10	2	3
4	6	5	8	9	7	10	4	1	1	2	5	4	3	6

sequence is divided into ten basic steps. For each step except Steps 1 and 10, four individual probe sheets have been created. The pinpointed behavior is *writes digits.* The first two probes written for each step in the sequence always contain combinations in forward or backward order, for example:

0   1   2

+0, +0, +0 ;   or

9   8   7

+0, +0, +0   . . .

in conjunction with the arithmetic program, which teaches the combinations using a number line. The third probe sheet presents that step's combinations in random order, for example:

6   2   9   5

+0, +0, +0, +0   . . .

For Steps 2 through 9, a fourth and final probe combines all the sums taught thus far in random order. Thus, the final probe sheet for Step 2–introducing sums of "+1"–contains both +0 and +1 combinations, in random order.

Probe Master Sequence: Add-Facts 0-9

1. *Writes digits* in answer to *+0* combinations: sums to 9; top addend varying between 0 and 9, bottom addend 0.

    a. Forward order, sums 0-9 (e.g., 0+0, 1+0, 2+0 . . .)
    b. Backward order, sums 9-0 (e.g., 9+0, 8+0, 7+0 . . .)
    c. Random order, sums 0-9.

2. *Writes digits* in answer to *+1* combinations: sums to 9; top addend varying between 0 and 9, bottom addend 1.

    a. Forward order.
    b. Backward order.
    c. Random order.
    d. Random combinations: +0, +1.

3. *Writes digits* in answer to *+2* combinations.

    a. Forward order.
    b. Backward order.
    c. Random order.
    d. Random combinations: +0, +1, +2.

4. *Writes digits* in answer to *+3* combinations.

    a. Forward order.
    b. Backward order.
    c. Random order.
    d. Random combinations: +0, +1, +2, +3.

    . . .

10. *Writes digits* in answer to *+9* combinations; sums to 9.

    a. Random combinations: +0 to +9.

All of the probes are designed with 80 combinations on a page, 10 in a horizontal row. The final probe sheet in the series (labeled Step 10a in the figure) summarizes all the problems presented for the entire sequence, random sums zero to 9.

Figure 3:9 shows the probe Mr. Archer always gives first. This probe is the last in the sequence the teacher has designed, and represents the terminal behavior Nigel hopes each child will have when instruction is completed. He presents it first, to rule out the possibility that a child has already learned all the desired skills. Also, because the probe contains samples of all possible combinations, the teacher can conduct an error analysis, to determine with which combinations a child seems to have the most difficulty.

Mr. Archer administers the probe as a one-minute sample, three days in a row, with a criterion of 30 digits per minute correct, zero errors, for two out of the three days. This criterion is based upon the work of pupils who have shown success in the past on the instructional sequence. If the child meets the criterion, he goes on to the next part of the instructional sequence. If he does not, Mr. Archer determines at what particular slice of the curriculum he should be functioning.

Step 10a: Random Addends 0-9

Name: _____

Date: _____

Time: _____

Count: say _____ / write _____

7	0	2	1	0	2	2	5	1	0
+2	+9	+5	+7	+9	+1	+7	+3	+5	+1
4	0	4	9	2	3	3	4	0	1
+2	+6	+4	+0	+3	+2	+3	+3	+9	+8
3	5	1	0	4	2	4	0	2	1
+5	+2	+6	+3	+0	+5	+5	+8	+2	+2
8	6	0	2	1	3	5	0	1	2
+0	+3	+5	+2	+2	+0	+4	+8	+1	+6
7	1	3	5	4	1	1	6	0	4
+0	+5	+3	+1	+5	+3	+8	+2	+2	+1
5	3	5	0	0	0	3	1	4	0
+3	+5	+0	+4	+0	+7	+6	+0	+4	+4
0	3	1	4	0	0	8	3	2	0
+0	+6	+0	+4	+4	+7	+1	+6	+3	+2
2	6	3	3	4	6	3	6	5	2
+6	+3	+1	+4	+3	+1	+4	+0	+4	+4

Archer chooses to use only worksheet "d" of each sequence step as a probe. The other worksheets, "a," "b," and "c," function as practice material in addition to other instructional materials on which the child works each day. Each child decides when he is "ready," and asks to take the probe. Since the instructional sequence is set up around a "criterion" rationale, no child progresses to a new step in the sequence until he has met the criterion on the present step. At the same time, a child is free to progress as quickly as he can, at his own pace, rather than waiting for the rest of the children to "master" the required material. In this manner, the child works forward until he again reaches the final step in the sequence.

Mr. Archer's probe set contains numerous individual probe sheets, each coded to the master sequence, and each designed to make scoring and recording as easy as possible. Although the set required an initial outlay of time and effort, Nigel Archer finds it well worth the work: the probe sheets serve a combination of initial assessment device, practice material, and criterion-referenced "retest" functions. The format encourages systematic progression through the instructional unit. Mr. Archer has used the same probe set for several years now, and shares it with colleagues who are equally pleased. Some of these colleagues have found Archer's probes to be extremely useful as an inventory device. They administer probes at the outset of an instructional unit on basic addition facts to determine how pupils are functioning *before* deciding what direction the program should take and what criteria might be most appropriate for each individual.

   Other probes taking Archer's format. Many skill areas lend themselves to a probe format similar to the one just described. Figure 3:10, for instance, illustrates a word recognition probe, sampling medial short-vowel (c-v-c) combinations. If the youngster experiences difficulty with this list, the teacher may decide to slice back in the program, requiring him to practice on a worksheet such as that in Figure 3:11, where a single vowel (e.g., short *u*) is concentrated upon.

FIGURE 3:10. *Medial Short Vowel Word Recognition Probe.*

Name: _____

Date: _____

Correct: _____

Error: _____

cut	bag	dig	jet	had	hit	gap	mud	pig	zip
not	red	pen	sod	pat	jog	tin	gun	pop	pan
mug	cap	bit	hut	bad	jig	lap	bet	hug	fan
hop	bun	fin	dog	cut	pod	hen	fed	dot	lip
fig	cud	mut	map	rag	kit	pig	lad	met	vat
tug	log	van	bin	cop	sun	bud	cod	dig	den
hip	bed	cot	rod	red	led	toy	win	nip	sat
dud	hog	tip	run	got	men	ran	hat	nut	tag
fog	pin	fig	mad	mop	pet	fun	man	pit	nap
lug	tap	wig	sit	wet	rug	sad	rut	sad	bin

Another teacher, in a classroom for moderately handicapped youngsters, might design probes based on the same model to determine which of the important "survival" signs frequently posted in public places her pupils were able to read. Using the idea of presenting a large number of like items on a page, she constructs a sheet containing 40 items, featuring ten of the most commonly seen outdoor signs (e.g., "Men," "Women," "Walk," "Stop," etc.). The signs are drawn in miniature, using the international pictograph system, and randomized on the page. The teacher asks the youngsters to point to the signs and say the word or concept they represent, "reading" as many as they are able in a one-minute period. This probe, like the others, can be easily "sliced back" to survey a smaller portion of the curriculum.

   These are only a few of the many hundreds of probes that can be designed according to this format. Required responses can be written, spoken, or signed. The teacher can sit beside the child and record responses as they occur, or, in the case of written responses, wait until a later, more con-

FIGURE 3:11. *Medial Short* u *Sound Word Recognition Probe.*

Name: _____

Date: _____

Correct: _____

Error: _____

gun	rum	fun	sun	bun	cup	hut	rub	tub	cub
but	fun	bus	sub	bum	rub	jut	gum	hum	hub
tub	cup	bud	rub	hut	hug	sun	run	sun	but
nut	gum	bum	cut	hut	jug	rug	tub	cub	jut
rut	bus	bun	bum	bud	fun	gum	hum	sum	pub
but	dug	tub	rug	jug	sun	run	sun	hum	cud
gun	but	cut	hut	jug	rug	tub	cub	mud	rut
but	cub	cud	cup	tub	gum	hum	sum	pub	cud
cud	cut	bum	sun	hut	sun	run	sun	hum	cup
pub	but	cut	hut	jug	rug	tub	cub	but	tub

venient time to correct and record corrects and errors.

Although the examples shown lend themselves readily to limited time/unlimited response samplings—generally a one-minute timing—they can easily be administered as fixed item/unlimited time probes. We remind you again that such probes can be given with preestablished performance requirements in mind, or criteria may be established after learning profiles on probes have been thoroughly examined.

Creative probe design. Probes can take a virtually unlimited number of shapes, covering an unlimited range of curriculum content. Consider these examples:

Probes for a Social Studies Ecology Unit
(Contributed by S. Gorman)

Figure 3:12 contains a list of probe items designed to assess various aspects of performance in an ecology-oriented social studies class for primary-age youngsters. Although the probes could have been devised in such a way as to set a required number of correct responses with no particular time limit, the teacher has established a time limit and left the children to give as many responses as they can within this period. Eventually she counts only those answers which are new responses for a child, showing that the child is expanding his thinking about the environment.

The teacher administers these probes several times during the course of the instructional sequence, counting the number of answers the child offers each day. She intends eventually to examine other behaviors—including verbal generalizations, predictive statements, related comments (those which relate to one another and the general topic of discussion), questions asked, number of types of questions asked, number of verbal relationships made between two events, objects, etc.—to sample more complex response patterns on the part of each child.

1. Builds a natural environment using a tray of dirt and a box of trees (miniature plastic), grass, rocks, blue rubber for water, flowers, animals, fish, insects (all plastic). After completed, names things used to make up a natural environment—one minute.
2. Names things that people can do in the natural environment—one minute.
3. Builds a city using a tray of dirt and a box of trees, flowers, cars, buildings, rubber for roads, bridge (miniature plastic)—ten minutes. After completed, names things used to make up a city—one minute.
4. Names things that people can do in the city—one minute.
5. Builds a city in a tray of dirt, etc., that has been set up for a natural environment—ten minutes. After completed, names changes of the natural environment as the city was built over it—one minute.
6. Names possible effects of these changes—one minute.
7. Names some problems of a city that gradually affect the natural environment—one minute.
8. Names some things that people can do to eliminate or lessen these problems—one minute.
9. If a man were on an island, name the things he would need to stay alive—one minute.
10. Names several things that man does not need just to stay alive—one minute.

Probes for a Geography Maps Unit
(Contributed by K. Wilson)

A special education teacher in a rural junior high school devised the probes in Figure 3:13 to test her ninth-grade pupils' abilities to draw simple maps and to read maps correctly, following the routes indicated on them.

The probes were designed after several frustrating attempts to introduce geography skills as a part of district requirements. Her pupils' inability to cope with such immediate applications as finding their way from school to home unassisted made more abstract concepts—(e.g., states, countries, geographic features such as river and mountain systems)—seem inappropriate. The items she came up with were an attempt to sample the most basic principles of map making and interpretation in relation to the surrounding community. Obviously, this teacher is more concerned with the number of items performed correctly than with the amount of time taken to perform.

After implementing the probes once and finding her pupils virtually unable to provide any correct responses, the teacher decided to stop assessment and initiate instruction toward the objectives that her newly designed probes made clear. During the course of instruction, she used the probes frequently as periodic progress checks—a system she found vastly favorable to her former strategy of testing at the end of complete instructional units.

The probes made the precise purpose and direction of instruction clear to all. The pupils were excited by the probe process which gave them an ongoing picture of their own improvement. They looked forward to the daily "map trips" eagerly; and, as they gained increasing skills, were able to satisfy educational objectives as well as developing a new confidence and sense of independence.

Objective: At the end of the term, the student will *demonstrate an understanding* of maps as symbols of geographical areas by completing the following tasks:

1. The student will draw a map showing the route between his house and school.

    Conditions: The map will be drawn with paper and pencil, without aids, and within five minutes.

    Criteria: The map will be scored by comparison with a teacher-drawn map of the same area, with one point for each correctly placed road, landmark, label, direction-finder, and indicated turn. Minimum acceptable performance will be at least four points, including a correct direction-finder.

2. Given a compass and a map of his city with starting place and destination marked, the student will direct a driver along the streets from start to destination.

    Conditions: Each student, riding with a small group in the teacher's car, will direct a course designed by the teacher to be commensurate with his ability, but not less than nine blocks distance and with at least one left or right turn.

    Criteria: To be judged successful, the student will issue complete instructions to the driver, including "start," "drive straight down this street," "turn right (or left) at the next intersection," and "stop here," each at the appropriate time from start to finish. One point will be scored for each correct direction, with a minimum of five points for acceptable performance.

CONCLUSION

This chapter has presented a range of evaluation strategies and devices. Some, such as the standardized tests, may be required procedures in order to meet certain program dictates. Of these, most will give you little information that can be applied directly to questions about likely performance within specific instructional materials. What information you do pull from standardized test results must always be examined critically in terms of its reliability and validity. Other evaluation strategies—those of an informal nature—will be largely up to your own choosing. Often you will find that you have to design these instruments for yourself. While informal tests have the capability of addressing your programming needs much more directly than can any of the standardized tests, it is still up to you to ensure that the devices you select or create are best suited to the child and to the material and skills you intend to assess.

Chapter 4
Creating Functional Educational Objectives

Goals and objectives figure informally in many areas of our lives from childhood on. As youngsters we cherish dreams of becoming doctors, bank presidents, scientists, athletes, and astronauts. When we grow older, our objectives generally, although not always, become more concrete and more close-ly linked to our actual capabilities and needs. Still, for most of us, setting objectives in our private lives is seldom a formal procedure. We tend to keep our plans loose enough and vague enough that it is easy to let them slide from one year to the next. Perhaps for this reason, some of us bridle at the idea of specifying formal objectives as a part of our teaching procedures. Others of us think, "Well, it's a good idea," but we have difficulty relating objectives we come up with to what actually goes on in our classroom instruction.

Over the past twenty years, objectives have become an increasingly established part of class-room reality. Many school systems insist that their teachers write out detailed behavioral objectives. For certain programs, funding is often contingent upon the specification of annual goals and objec-tives for each pupil in the program. And yet, in too many instances, objectives remain a formality with no real connection to instruction. Teachers may engage in formulating objectives; but, all too frequently, this laborious exercise is the end. Instruction itself does not necessarily focus on the specific behaviors defined in the objective, nor are targeted behaviors monitored throughout the in-structional period to learn whether or not the objective is being reached. Seen in such a light, objec-tives are truly paper lions—menacing, but with no real bite to them.

This chapter is intended to help you turn those "phantom objectives," as one educator called them (Lovitt 1980), into the real thing. Writing objectives is an operation that can be simple and quick. But more important than making objective-writing easy, we intend to make it *functional*. Far from being a mere formality, an objective is a concrete statement of your intention. It is an initial, and important, statement of teaching commitment. Each objective is a bridge between what you have determined to be the child's present capabilities and needs, and the educational program you are capable of providing.

DEFINING AN EDUCATIONAL NEED

The first step in identifying an objective is to construct a statement of educational need. Such a statement reflects the interaction of teacher expectations and present pupil performance. A simplified equation can be drawn which represents this interaction:

Instructional Expectation – Learner's Current Status = Educational Need

As the equation indicates, in order to come up with a functional educational objective, you must define, first, your own instructional expectations, and, second, the child's current performance capabilities. "Subtracting" learner capabilities from instructional expectations and requirements, you are left with a statement of educational need on the part of the pupil. It is toward this need that each objective is addressed.

Major Components in the Learning Equation

Instructional expectations. Identifying and ordering the forces that influence teaching has already been discussed in Chapter 2. To review briefly, instructional expectations and priorities are determined by a complex set of factors, including the administrative or program requirements to which you must respond, the demands of the instructional materials you'll be using, and the dictates of curriculum guidelines, as well as your own training, teaching philosophies, and areas of instructional expertise or interest. These factors combine to determine the instructional material you will stress, as well as the instructional procedures you adopt. It is against this combination of content and approach that the child's entry performance is compared and evaluated.

Learner's current status and arriving at a statement of need. Your evaluation of the child's current performance can be carried out in a number of ways, including referral to reports and observations made by others interested in the child, informal direct observation, and/or the use of any of the assessment devices discussed in Chapter 3. Try to summarize the results of your evaluation efforts in terms of areas of need, relating needs to categories which you find make sense to your own program, and, at the same time, give a broad view of the child in question. For example, using a developmental framework for a young or developmentally delayed child, you might consider performance in terms of areas labeled cognitive, language, self-help, social, fine- and gross-motor, and so on.

- Identify the child's strong areas of performance (e.g., gross-motor, social).
- Identify the child's weak areas (e.g., math, reading).
- Determine whether or not a pattern of needs and capabilities can be identified. Do needs seem to cluster in one or two areas?
- Within each of the areas examined, can you pinpoint specific skills that require special attention (e.g., math: time-telling, fractions, two-column addition problems involving carrying)?
- Reexamine both strong and weak areas and related skills you have identified. Can you rank order the weak areas in terms of any guidelines your assessment tool or your own observations might suggest? Can you rank order, as well, the areas of performance strength? It is often important for the child's sake, rather than concentrate solely on areas of skill weakness, for you to include as a part of a total program an area where he demonstrates relative success, adding to and refining these skills.

Decide which objectives are important, and discard those that aren't. Objectives can get away from you. You may start out with one or two strong ones, and then, to your dismay, they multiply; and before you know it, they're out of control. Where objectives are concerned, you are much better off with only a few that you *know* you will use to guide the development of instructional plans and to structure ongoing assessment of pupil progress. We cannot stress too strongly that *the only good objective is a functional objective.* If you don't intend to use an objective, get rid of it!

The priorities outlined in Chapter 2 act as one kind of framework to use in deciding which educational needs merit the effort needed to turn them into a formal objectives statement.

Do a cost analysis. Comparing the "cost" of an objective—in terms of instructional hours spent by both yourself and the pupil—with the benefits that will result, is another way of keeping objectives within the bounds of reality. If a sixteen-year-old handicapped youngster with no reading skills is referred to your classroom, you may decide to take him through a traditional reading program, specifying that by the end of the year he will read with 90 percent accuracy from a basal primer text. But consider the hours each day you and that youngster will spend developing such skills, and with what results? Does reading a primer help this individual function with greater ease in the world outside the classroom? The "costs" in this case are greater than the benefits. A more appropriate objective might be to teach the same youngster to identify with 100 percent accuracy, survival vocabulary (e.g., Men, Women, Danger, Exit), can labels, and food preparation instructions that will enable him to participate more fully and capably in independent living activities.

Create an orderly sequence of objectives. Another way to keep your objectives under control is to arrange them in terms of some kind of sequence. Current educational trends require teachers to define both short- and long-term objectives, implying that all instructional tasks can be broken down into intermediate (short-term) components which can be related to a final (long-term) target.

— *Long-term objectives.* These represent what a given child will be able to do at the end of a certain instructional sequence. If they represent one-year time periods, they are sometimes referred to as "annual goals." They may represent longer periods—for instance, biannual—as well, especially in situations such as a special education classroom, where certain goals can easily require longer than a year to complete.

  Long-term goals must be global enough to encompass specific short-term objectives, yet stated with sufficient detail to make them easily communicated to all persons interested in the child's progress. Such goals are often based upon a "best guesstimate" of where a child's performance will be at the end of a specified period.

— *Short-term objectives.* These are a sequenced set of objectives describing the steps the child must master as he progresses from his current level of performance toward the long-term objective defined for him.

A long-term objective, and the short-term objectives leading to it, are frequently identified in relation to behavioral checklists, curriculum sequences, and administrative guidelines. Sometimes, however, rather than resorting to such frameworks, teachers rely on *task analysis.* This is a process whereby the task itself is broken down into sequential components. Task analysis involves isolating, describing, and sequencing all of a specified task's major subtasks. Such a breakout assumes that, when all necessary subtasks have been mastered, the child will automatically be able to perform the terminal behavior, or long-term objective.

Subtasks can be isolated, described, and sequenced in a number of ways, including:

— *Reverse analysis*, starting with the long-term objective, deciding what step immediately precedes that, then what steps precede each intermediate step until a beginning level of performance is arrived at.

- *Forward analysis,* working step-by-step from the child's present skill level to an eventual long-term goal.

To experience for yourself the process of task analysis, imagine that you are teaching a developmentally delayed youngster to button one button on a shirt. Close your eyes and envision the steps, from beginning to end, that he must master in order to acquire what seems an ordinary and easy task. The sequence might look something like that outlined in Figure 4:1.

FIGURE 4:1. *Task Analysis: Buttoning.*

---

1. Grasp button with pincer grasp.
2. Grasp material surrounding opposing buttonhole using pincer grasp with opposite hand.
3. Pull button and opposing buttonhole material together so that they almost touch.
4. Align button with buttonhole.
5. Turn material with buttonhole outward as though to start buttoning motion.
6. Push button halfway through hole.
7. Touch thumb of opposing hand to button and release material, freeing forefinger of opposing hand to pull button through.
8. Grasp button with opposing hand, releasing grasp with initial hand.
9. Pull button through hole.

The sequence thus identified is neither developmental nor definitive in nature, but, rather, dependent upon the functional requirements and the basic structure of the task itself. How many steps are identified, and in what detail they are described, will depend upon the learning abilities and requirements of the children with whom you work. Youngsters demonstrating significant learning problems will most likely benefit from minute step breakouts. Each of these steps may require special programming, and several blocks of instructional time for certain youngsters.

DETERMINING THE COMPONENTS OF EDUCATIONAL OBJECTIVES

Various educational sources provide examples of ready-made objectives. You'll note that these objectives are written in a number of ways. Some sources present objectives with no mention of criteria, while others specify criteria in the utmost detail. Some describe the conditions under which performance is expected, and others leave such conditions as a matter of speculation. All, however, share two characteristics: (1) a clear identification of desired behavior; and (2) description of the curriculum content in the context of which behavior is expected to occur.

Behavior and content are mandatory components of every objective. Other components described in the following material may be included or not at your discretion. Each allows you to describe an objective with increased precision, offering definite guidelines by which to plan and evaluate the instructional program. If you choose to include the additional components, you may use or exclude whichever you wish. We have presented them within a certain order here, simply for organizational purposes; but that order should not imply that the components are sequential or cumulative.

Pinpointing Behavior

The basis of any clear functional objective is the identification of the behavior you expect to observe. What must the child *do* to indicate to you that he has understood and mastered the instructional sequence you have identified? "*To write* answers to basic addition facts"; "*to write* formulas for three common cleaning solutions"; and "*to write* an essay describing the significance of 'light' imagery in the play, *Hamlet*" are three very different objectives for which a behavior—*to write*—has been specified. Obviously, while behavior forms the core of the objective, it cannot stand alone as a statement of educational need. You must also specify *content*; that is, *what* it is the child does.

Describing Content

The three examples just given shared a common behavior—to write. What differentiated these objectives was the content associated with the behavior in each case, (i.e., answers to addition facts; chemical formulas; an essay). Consider the following points as you specify behavioral content:

Level of cognitive performance. What level of cognitive performance are you requiring? Are you confining objectives to fairly simple operations such as demanding only recognition or recall? For example, "To write the appropriate date of the discovery of gold in California," involves a much different behavioral content than an objective which asks the child "To write an essay clarifying the significance of the discovery of gold in California." One asks the youngster to write (behavior) a single *date* (content), while the other requires that he write (behavior) an *essay* (content). Where it is appropriate, don't be content to settle for objectives directed toward the lowest levels of cognitive functioning. Regardless of the instructional material used, aim toward the higher levels of cognitive hierarchy.

Content generality. To test your objectives for content generality, ask whether they apply to more than a single lesson or a single test item. Each short-term objective you write should represent an instructional unit you have planned, rather than one instructional session. Long-term objectives can, of course, cover periods of one year and more, and are the outgrowth of several units.

Teachers in regular classroom situations, trying for greater content generality in their objectives, might formulate objectives such as those in the right-hand list in Figure 4:2 which, in contrast to the objectives listed on the left, are more general in content.

FIGURE 4:2. *Giving Objectives Content Generality.*

Specific	*General*
Demonstrate how to do a sit-up.	Demonstrate any of six basic warm-up exercises.
State in writing the conflict in *Wuthering Heights.*	Define the tone of a 19th century Gothic novel.
List the economic effects of acquiring Alaska.	List the effects of any territorial expansion by the United States prior to 1900.

Content limits. Carefully define the content class to which the objective is relevant. Clear content limits provide the basis for selecting appropriate items and materials. Such limits are especially helpful if they can be drawn to include the range of content the child will be expected to master.

Try to identify a particular class of problems or problem types rather than the specific problems themselves. As you attempt to identify groups of problems, look for similarities such as units of instruction, topics, or issues, that have similar intent. Include specific details which further define the range of content you have in mind, as they apply to a particular class of problems (e.g., "Writes digits in answer to *two-column addition problems, integers ranging from 10 to 99*"; "Reads aloud words using a *c-v-c pattern in which vowels include a, e, and i*"; "Buttons buttons, *ranging in size from one-half inch to 1½ inches*").

Specifying Conditions

It is important to describe performance conditions, if you think there is something about the conditions under which you expect the child to behave which may be critical to his success or failure. In other words, were conditions different, would the youngster be more likely to achieve the objective; less likely? This is an especially important issue for many handicapped youngsters for whom generalization is often a problem. One teacher might report that Billy has mastered a certain set of behaviors, while another teacher may fail to find any evidence of his success. The difference between the two reports might be accounted for by the conditions under which they expected performance. Factors such as working on a one-to-one basis with the child, presenting materials on flashcards versus on a blackboard, presenting 10 rather than 50 items at a time, allowing the child to work in the classroom on his assignment rather than in a separate area with which he is not familiar—each, or all, could conceivably contribute to the disparity in the results observed.

The conditions you may need to consider include:

How. How is the program to be presented? Examine the details of your program. Is there anything about your presentation which could influence performance of the desired behavior? If a change were to be made in certain of those details, might the child perform differently? Are instructions to be given? How will they be delivered—with gestures, words, pictures; will there be a demonstration? Is the child given a certain number of opportunities to respond? Is there any choice in the matter? Objectives requiring specifications for "when given," "shown," or "asked" all fall into this category. Determine just how specific you will need to be. The following represent various degrees of specificity:

— *"Shown four items, with three having the same size, shape, or color*, the child will point to one which is 'not the same.' "
— *"Given pictures in random order*, the child will place the pictures in a story telling sequence."
— *"Without the aid of dictionaries*, the child will show correct word pronunciation, using phonetic markings."

What. What type of material you actually plan to present could also be a major factor in performance. Does this material have any special distinguishing features (e.g., size, shape, texture, color, auditory properties or reinforcing value)?

— *"Using felt tip marking pens*, the child will write cursive letters Aa-Zz."
— *"Using containers of water*, the child will demonstrate the mathematical properties of 'conservation.' "

Where. Is there anything significant about where the child is to perform; will this place influence behavior in any way, making it worth mention when specifying the objective (e.g., standing next to the coat rack, sitting next to the snack table, sitting next to the teacher at lunch, standing at the blackboard, standing in front of the class, on a crowded playground or in a special room with just the child and an adult present)?

    — *"Sitting at the lunch table*, the child will answer questions directed to him."

When. Is the time when you would expect the child to perform a significant factor? For instance, are stipulations such as *morning versus afternoon, before lunch, during reading period, during recess, at the end of the day, at the end of the term* crucial? Might the child perform differently if behavior is expected during his daily routine (as he eats, dresses himself, plays on the playground), as opposed to isolating a particular program time in which to concentrate on the behavior of concern?

    — *"During morning recess*, the child will approach and make comments to peers."

Who. Will the person involved in presenting or carrying out various aspects of the program with the child influence performance (e.g., significant people to consider include parents, counselors, principal, volunteers, teacher's aide)? Will it be necessary for anyone else to be present? Also, is the child in any way actively involved in the presentation; that is, will he have any control over getting or presenting material himself?

    — *"Accompanied by the school librarian*, the child will select books for free reading."

Setting a Criterion Level

Adding a criterion level component to an objective implies the question, "By what standards will I judge that the child has, in fact, achieved my objective for him?" Among the steps for defining a performance criterion are these:

Step 1: Deciding what feature of performance is most important. The first essential step to consider is how you are planning to observe a child's performance and how you will know if the child has made progress or not. Some unit of measurement must be selected at this point. Factors in your decision might include:

    — Is time the big factor? Would you like him to spend more or less time engaged in the behavior? Would you like him to respond more quickly once the cue is given to him?
        "The child should sit at his desk *for 20 minutes*,"
        "The child should say answers *within 5 seconds* for each flashcard containing words from the Dolch sight words list."
    — Is number of responses the major factor?
        "The child will approach and make comments to peers *five times each recess period*."
        "The child will ask *ten questions* each science period."
    — Is accuracy the major factor?
        "The child will read words orally *with 95 percent accuracy* from his Grade 1 text."
        "Given word problems at his grade level, the child will provide solutions *with 98 percent accuracy*."
    — Is speed, as well as accuracy, the issue?
        "Given a text at second-grade level, the child will read orally *100 words per minute with no more than two errors per minute*."

"The child will write digits in answer to basic addition facts at *30 per minute with errors not to exceed one per minute.*"

Step 2: Selecting a desired level of performance. In addition to knowing what feature of performance (e.g., accuracy, speed, number of responses) is most important to you, it is also helpful to specify at what level you expect that performance feature to be displayed. For instance, in one of the objectives just cited, specifying accuracy as a crucial feature, the teacher further qualified the criterion by stating that accuracy must be at 98 percent. How do you go about deciding at what criterion level you will require performance? At this time, there is no one best way to establish criterion levels. Perhaps some day in the near future, a great deal more data will be available from which to draw this information. Until then, the following seem to represent the most common alternatives available to us in establishing these levels.

— *Consult established authorities.* For every instructional area, there are numerous authorities whose books, papers, and lectures can offer concrete guidelines regarding criterion expectations. Read, listen, and ask questions with the intent of finding out such information. Corner any convenient resource person to whom you have access (e.g., the reading or math specialist in your school, or district coordinators for your area of interest). Consult appropriate curriculum guides, and, of course, whatever district or state administrative guidelines that might be applicable or take priority for your program.

— *Normative data available.* Some authorities will offer their own experienced opinion, while others may cite actual data. You may want to try looking for the results of actual studies applicable to the question you have. For example, developmental guidelines based upon sampling the performances of large populations can provide a helpful perspective as you attempt to establish criterion levels in areas such as early language, motor, social, self-help, and preacademic skills. Consult books and journals directly applicable to the area of interest in order to locate studies which have involved collecting data for a large number of children on the same skill or topic.

Whenever you use such data, however, remember that these are only guidelines, not hard and fast numbers. The child for whom you are trying to establish criterion levels may differ significantly from the sample population in respect to which original performance criteria have been determined. You will need to take into account the individual youngster's background and performance characteristics, comparing these to those of the sample population and making whatever adjustments are necessary.

— *Consider child's previous performance.* One popular way to establish criterion levels for an individual is to use information about his previous performance. If the child has ever engaged in any program similar to the one you plan, check whether data may be available. Consider the child whose previous reported level of accuracy was 50 percent in his grade level text. It would be unreasonable to immediately establish 100 percent as a criterion level for this year, unless drastic changes in curriculum are planned, such as slicing back to earlier levels where greater success is possible initially.

— *Consider child's present performance at program entry.* The child's entry level performance can be determined through the assessment procedures already described, or simply by monitoring day-to-day performance levels over a brief period of time. A criterion-referenced test will, of course, provide built-in guidelines concerning eventual performance goals. The child's response in comparison to such guidelines will give you a better idea of whether or not such criteria are valid for him. Rather than administer a criterion-referenced test, you may decide to use an inventory or a set of probes, deciding later, based upon the child's performance, what specific criteria should apply.

— *Consider peer performance.* Another increasingly popular way to establish criterion levels that suit both the demands of the objective and the demonstrated performance characteristics of a particular child, is the peer comparison method. Other youngsters—often those who have already mastered the task—are used as models, against whose performance that of the child in question can be compared. Peers may either be considered in a "comparable" population (e.g., in terms of handicap or educational setting), or may be those already functioning within the setting you desire this child to enter. Examining peer performance is often the easiest and most concrete way of gathering performance expectation information.

Suppose, for example, you are a special education teacher who is trying to get one one of your youngsters back into a regular classroom setting. You might ask the regular education teacher with whom you are seeking to make the arrangement if you can gather a bit of data on some students she considers in the "high," "average," and "low" groups. Setting up probes for a few of your major objectives, you might try to arrange to collect data in the least time consuming way possible (e.g., using probes involving one-minute samples). With the results of these probes, you can determine levels of expected performance for each of the three designated groups. You now have a concrete reference for criterion levels that refer to the real performance of children within the setting your youngster hopes eventually to enter. (Further information regarding the use of peer performance to set performance objectives is found in Chapter 12.)

Peer comparison is also a useful strategy for examining social (or antisocial) behavior. Assume you are interested in knowing how often a child will be allowed to engage in behaviors such as talking-out without being given permission by his teacher, getting out of his seat, or interrupting—or how frequently he will be expected to ask questions; to offer to share his possessions with others. Different teachers have different criteria and tolerate various levels of performance, and it is very helpful to know just what is tolerated by the teacher to whom you wish to refer the child. This will give you a clear idea of where you might aim your objectives to meet teacher expectations in that particular setting.

You may even want to analyze peer performance in terms of a basic behavior needed to perform the objective. Using an arithmetic example, suppose the peer were to write digits at 50 per minute; yet, when he was given an assignment requiring him to provide answers to basic addition combinations, his rate dropped to 25 per minute. You might then establish a criterion level for the child in question by using his basic behavior data. If he writes digits at 10 per minute, his rate, too, might drop by half; therefore, he might be expected to write answers to basic math combinations at a rate of 5 per minute.

— *Transfer and generalization.* One of the biggest instructional questions concerns what the child will do with his training; that is, how he will apply it in the world outside the classroom. For handicapped youngsters, especially, this is an important issue. What level of performance must a child reach before he is able to generalize? How many times, for instance, should he demonstrate success buttoning within the structure of a program before he will spontaneously button any button on his own clothing, in or out of the structured setting? The same kinds of questions can be asked about change-making, time-telling, or any other instructional content we present. How successful must the child be to apply his skills successfully, and *independently*, elsewhere?

A related issue concerns performance on sequential steps in the curriculum, or, more generally, to what degree a child must master a skill in order for that skill to be useful to him in future achievement. An example is the teacher who was interested in oral reading and accepted 80 words per minute read orally with six or fewer errors as a criterion for moving on to the next story in a reading text. She found that the data from the next few days in the following story had a median score of 50 words per minute with 18 errors per minute. In another, comparable, text, she set 100 words per minute, with two or fewer errors as a criterion, with data this time showing that the child performed in the next story with a median of 90 words per minute and four errors. The teacher's data indicate some important information about setting criterion levels that are most functional for the individual child involved.

If you are interested in exploring functional levels of performance, you must consider how to set up and periodically examine those situations to which you want transfer and/or generalization to occur. You may find in asking these questions that your program itself takes on new dimensions. One geography teacher started arranging regular field trips to administer probes, another involved with public speaking arranged real-life speaking engagements for her pupils; a composition teacher required letters to real people, to name only a few examples.

— *Ceiling performance.* Sometimes the way you structure the program will, in itself, be a major factor in determining the criterion level. If you give 50 flashcards, the child can't get any more than 50 correct; presenting a page with 25 problems sets an automatic ceiling at 25; asking six comprehension questions locks the child in at a maximum performance of six correct. The mode of program presentation may also be critical in determining ceiling performance. For instance, a mechanical device such as Language Master limits the child's performance in terms of the number of cards he can put through the machine in the amount of time allowed. In an example considered in the last chapter, dictating spelling words and waiting for every child in the class to finish writing before presenting the next word, establishes a performance ceiling.

Keep in mind, too, the basic movement cycles that a program requires. A child is limited by his skills in the basic movements such as writing digits, letters, pointing or turning pages. If he has problems with these, his problems will only be compounded when given the added requirement of trying to answer questions or obtain information using these movements. Consider the child who writes digits at 10 per minute when given a page of random digits to simply copy. Surely his rate will drop when he is confronted with a task requiring that he write numbers in answer to various problems given him. To set a criterion of 30 per minute for arithmetic tasks would be unreasonable at this point.

Step 3: Deciding what time constraints apply. Consider how time figures as a factor in establishing a level of desired performance. As with every other type of criterion-statement, the time element considered crucial must be appropriate to the particular student for whom the objective is intended.

— Determine whether the level of behavior you specify must occur *within any particular time period* in order to fully satisfy performance criterion. If such specification is deemed necessary, it can be included in the objective as follows:

"Computes 60 addition facts with 100 percent accuracy *within one-minute sample.*"
"Decelerates nail bites so they do not exceed *one bite for entire waking day.*"

— Determine whether the level of behavior must be *maintained over any particular period* to meet criterion. Sometimes performance of a stated objective *just once* is not considered sufficient to signify that the child has mastered the desired objective. For instance:

"Given a set of 20 spelling words with c-v-c pattern, the child will complete with 100 percent accuracy *for two days.*" (Perhaps the criterion even specifies that performance must occur over two *consecutive* days.)

Establishing Expectations Regarding Objective Completion

Inclusion of an expected or predicted completion date further clarifies an objective (e.g., "Given nine objects to classify, the child will be able to place objects into three appropriate groups with 100 percent accuracy by March 18, 1981"). Although it may seem extremely difficult to predict when an objective will be met at the outset of a program, it is important to give the time component at least some initial consideration, for we all operate within certain time constraints. While there are many uncertainties to deal with in regard to program time, some things can be counted upon. Time passes, and the child gets older. How long will he be in your classroom—one year; two years? These are the outside boundaries for objective completion. Holidays, interim school vacations will also figure in program timing. How much time can you realistically afford to devote to a particular objective?

It is important to sit down with your calendar as you are establishing objectives and try to determine, even if this is at best, tentatively, how much time you intend to devote to the achievement of an objective. This date can be altered as the child's performance suggests; but the date often serves as an important reference point in program planning. With a date for objective completion clearly in mind, you can modify classroom strategies as the time line warrants, keeping up efforts in certain areas which seem to be falling below expectations, while modifying plans or even including extra skills where the child gets ahead of your predictions. As you develop a series of short-term objectives designed to lead to a major long-term objective, these date specifications will be especially important in helping to ensure progress that is on target in relation to the total program plan which has been established.

TYING OBJECTIVE COMPONENTS TOGETHER

The components outlined in this chapter are all important to consider whenever you begin writing objectives. While two of these components (behavior and content) are required, there is a great deal of variability in what you will see in practice, and in what you will find necessary for your own purposes. None of the other components discussed—conditions, criterion level, and completion date—is required, but, as you can see in the objectives presented below, each adds an additional level of refinement and direction to the program you plan for the child:

— *Behavior/Content*:
    "The child will list synonyms to define specified words."
— *Conditions/Behavior/Content*:
    "Given a group of 10 words taken from his grade-level text, the child will list synonyms to define specified words."
— *Conditions/Behavior/Content/Criterion*:
    "Given a group of 10 words taken from his grade level text, the child will list synonyms to define words with 100 percent accuracy, by October 20, 1986.
— *Conditions/Behavior/Content/Criterion/Expectation Date*:
    "Given a group of 10 words taken from his grade level text, the child will list synonyms to define words with 100 percent accuracy, by October 20, 1983.

Some of us hesitate to tie up our instructional intentions in so small and tidy a package. After all, learning is a process and each child is an individual who engages in and emerges from that process in a unique way. We are not suggesting, however, that objectives prescribe how skills will be taught, nor how a child will interact with the materials and concepts he encounters. The roads we take as teachers—and the roads the children take as learners—are a matter of choice. An objective simply describes the territory into which we hope learning will lead teacher and pupil alike.

At its worst, setting objectives is an empty exercise; after it's done, we can file the results away somewhere and breathe a sigh of relief. At its best, defining educational objectives is an enterprise which provides the basis for the entire instructional program that follows.

SECTION 2
MEASURING BEHAVIOR

Introduction

One of our favorite teaching stories involves a beleaguered citizen who, in exchange for his life, promised that, before a year was out, he would teach the king's horse to talk. The man's best friend was horrified: "How could you promise such a thing! You'll never teach a horse to talk!" "Don't worry," the man replied. "A year is a long time. Before it's over, almost anything could happen. I may die, or the king may die; or the horse may die; or the horse *may* talk."

The stakes being what they are in teaching, most of us feel hesitant about sitting around an entire year waiting to see whether or not the horse will talk. We work extremely hard on a daily basis to make sure that the children in our classrooms will acquire the behaviors we feel are essential. Yet, how thoroughly and how often do we really check to see whether or not a child is progressing in the right direction, and whether or not the teaching strategies we have decided to employ are actually working?

Section 2, Measuring Behavior, takes you through the second step in the measurement process: finding a practical and efficient way to monitor the behavior you have identified as an indicator of pupil progress. As you read the material presented, keep in mind that the measures of performance described can be implemented in a number of ways.

"After" Measures of Performance

Some teachers prefer to work on an intuitive level. They incorporate feelings about the child's performance in their instructional decisions without making arrangements to collect concrete behavioral information relating to progress. These teachers may wait until the very end of a teaching unit to implement a check of the child's achievement. This is what might be called a "one-shot" measure of performance. The child either "gets it" or he doesn't. For the child who hasn't achieved, it may be too late; the opportunity to intervene has already passed.

"Before-and-After" Measures of Performance

A more effective way to measure behavior is the "Before-and-After" approach, for which the teacher conducts an initial assessment to determine a child's strengths or weaknesses upon program entry and follows up at the end of instruction by measuring final skill achievement. This method results in a more balanced picture of where the child started and the progress made over the course of instruction, but it still fails to address problems and needs that may be occurring throughout the course of instruction itself. It fails also to account for factors that may be responsible for progress or the lack of it during this period.

"Before-During-and-After" Measures of Performance

Educators are increasingly realizing the importance of intervening before it's too late. Rather than waiting around to see whether or not the child fails or succeeds by the skin of his teeth, they admit the necessity of looking at a child while he's engaged in the process of learning a behavior. This has led to an entirely new way of talking about behavior measurement. Rather than checking only in a formal, time-consuming testing situation, teachers are finding that quick checks on performance, which can be incorporated within the learning task itself (or built around occurrence of the behavior they would like to see change), are the most meaningful kind. The more frequent the checks, the better. For this reason, we stress throughout this book the value of frequent measurement, conducted, as much as possible, on a daily basis.

Once teachers learn how to make these informal performance samples, they discover that they are able to gather optimal information about a child's progress with minimal interruption of their normal daily classroom routine. The two chapters that follow are designed to help you devise measurement strategies closely tailored to your own classroom needs. The ultimate objective of this section is to make measurement an activity so easy and so natural, you will want to incorporate it as frequently as possible in your daily teaching activities. Teachers who go this second measurement mile rarely want to turn back. They find the information they are able to gather well worth the additional effort they expend.

Chapter 5
Measures of Behavior

All behaviors bear certain characteristics and dimensions in common. Regardless of whether you choose to look at a large, vaguely defined chunk, or at a precisely defined movement cycle—whether you are concerned with academic behavior, or with social—the behavior you pinpoint exhibits such characteristics as *duration, frequency, quantity,* and *accuracy*. These are only a few of the performance dimensions which go together to make up each behavior a child demonstrates. Such dimensions give a behavior its shape and identity; and the degree to which each, or a combination, of these characteristics dominates performance, determines whether a particular behavior helps a child succeed in the world or ultimately raises problems for him.

Even though each of a variety of performance dimensions is present in every behavior, you are not always interested in each dimension equally. If you want a child to stay in his seat longer, the behavior's duration is an important issue. If you want to help a youngster improve the quality of his math performance, then accuracy may be your chief concern. If you are trying to help a youngster in your typing class get a summer job, speed combined with accuracy will be important.

There are many ways of analysing a child's performance to determine whether or not it contains the right amount of those characteristics you consider important to success. This kind of analysis is called behavior measurement. You can think of it as a process much like taking a picture of the child's performance. For each of the facets of performance that might be important to you, there is a specific behavior measure which is best suited to producing the kind of picture you have in mind.

WHAT'S IN A MEASURE?

This chapter examines six different behavior measures. To understand how they differ from one another, think in terms of changing lenses on your favorite camera: what type of lens you select, as well as the way you adjust it, dictates the final shape and focus of the picture you take. If your lens is a wide-angle, your photograph will cover a large area, allowing a broad sweep of background, but not much detail. A telephoto, on the other hand, brings you in for a close-up of a single, narrowly focused subject. Each new lens changes your picture's final perspective. Even if you keep the photograph's subject identical from one picture to the next, the way in which you see it varies immensely, depending upon the perspective through which it has been viewed.

Behavioral perspectives change in much the same way, depending upon the behavioral measure you select. Each measure is like an adjustment of the lens, forcing you to focus upon a slightly different aspect of performance. Think of yourself as instructing your behavior to pose. You snap a picture, first from one angle, then from another: wide-angle, medium shot, close-up. Your model remains the same, but how different he appears from one picture to the next!

Bringing Behavior into Focus.

Sometimes we are conscious of how the behavior measure we choose shapes our perspective of behavior. At other times, we snap the shutter without realizing the implications a measure holds, and the part it plays in the resulting picture of performance. Ultimately, we may be confused—even bitterly disappointed—by the picture of behavior that develops. We find that the measure we have selected simply doesn't reflect a particular feature of the youngster's performance we are especially eager to examine; or we're just sure, at a gut level, that performance is changing, and yet we don't see these changes reflected in our data.

The purpose of the following material is to bring into focus the unique perspective each of the common measures of performance offers, and to highlight the advantages and disadvantages each entails. Each behavior measure is described according to a breakout which includes a definition, directions for calculating the measure, arrangements that are necessary to take data using the measure, and what we call the measure's "special focus." This last section details the specific performance aspects or characteristics a particular measure is most sensitive to. Following "Special Focus," "Limitations" outlines certain restrictions the measure might impose upon a total picture of performance. These limitations are described, not to discourage you from using any of the measures, but merely to make you aware of the ways in which a measure might obscure issues that are important to you. A final section, entitled "Making the Most of the Measure," describes strategies for standardizing data collection to make the final performance picture as closely reflective of pupil change as possible. Our intent is not to influence your choice of a measure, but rather, to inform that choice. You must select the behavior measure that is best for you based upon what you want to know about the performance of the youngsters with whom you work. Choosing the right lens and figuring the final focusing of your behavioral camera are entirely up to you.

An Old Standby: Grades

Because *measuring* performance implies a scientific approach, you may question the inclusion of grades in this chapter. But grades are so commonly used—whether out of necessity or out of choice—that they demand at least a nodding glance. In fact, since using grades may be a required part of your classroom routine, you need to know as much about them as possible.

Definition. Dictionaries usually include these elements in the definition of grades as they are used in education:

— A class of things that are the same stage or degree
— A position in a scale of ranks or qualities
— A mark indicating a degree of accomplishment in school

Calculating grades. Most grading systems consist of five letters—A, B, C, D, and E—given on individual assignments, then totaled and averaged at the end of a report period to produce a summary grade for the specified performance area. Sometimes, rather than letter marks, a grading system applies "High," "Medium," and "Low," or "Satisfactory," "Needs Improvement," and "Unsatisfactory" ranking labels.

Data collection arrangements. Grades do not require any special data collection arrangements. What arrangements and materials are necessary are up to the teacher, who usually writes letter grades on individual work, records them in a grade book, and transfers final grades to a student report form at the term's conclusion.

Special focus. Grades have become essentially a part of the classroom culture, as well as "de facto" reinforcers in the school and at home. They describe performance in terms of broad categories whose value is determined by the individual who assigns them.

Limitations. Despite their common application to classroom descriptions of performance, grades involve serious limitations.

— *No uniform standards for assignment.* Because each grade is supposed to represent a certain "class" of performance, it is often incorrectly assumed that youngsters who earn the same grade are all functioning at the same level. Criteria for grades are likely to vary dramatically. A single teacher may set and maintain precise, consistent grading standards within his own classroom; but, beyond that, the system breaks down. There can be no really fair comparison of grades across teachers and subject areas.

Exactly the same grade can have a variety of different meanings, depending upon the teacher who assigns it, as well as upon the nature of the material being taught. While one teacher may give an A to a child based upon his improvement, another looks at the bell shaped curve; or perhaps neatness, attitude, or how fast the assignment was completed. In other words, an innumerable set of variables may determine the final grade decision; therefore, to say that one youngster is an A student and another a C student, is misleading in any but the vaguest sense.

— *Grades describe performance imprecisely.* If you consider the immense differences among human beings, it is almost impossible to imagine how human performance can be summed up in just five letter grades. (Some scales are even more limited—for instance, an elementary school report card on which teachers are asked to rate performance in both social and academic areas as "Satisfactory," or "Needs Improvement.") In this sense, grades are a very gross measure of performance, attempting to express all the variability from one child to the next—or the changing performance of just one child from one point in time to the next—on a scale with an extremely limited number of increments. The introduction of "+'s" and "−'s" to some grading scales points up the failing of the system, and even beefed up in this way, the grading scale is not very descriptive.

The system is capable of indicating only comparative performance positions among students, and is insensitive, on the whole, to individual performance changes. If a child is learning, some sort of significant changes should be taking place. The kinds of changes learning affects in behavior, and how long it takes for a child to learn something completely new to him, are among the most important pieces of information we can obtain about a child, and yet these dimensions of performance are most often lost when grades are used as a measure.

Consider an example which may be more than painfully familiar to you. Two teachers enroll in a graduate course. One knows most of the material before she enters the class and, as a result, hardly cracks a book all quarter, but earns an A just the same. The other student finds the coursework new and difficult. The materials assigned seem like a foreign language to him, but he struggles to learn the terminology and concepts he encounters. At the end of the quarter, his efforts are summarized with a C. Although the grades in question may represent satisfaction of carefully devised criteria, they do not represent fairly the tremendous change and growth in the performance of one student, nor the failure to reach beyond former boundaries on the part of the other.

— *Grades promote labeling.* Perhaps most distressing is the fact that when we use grades, we often start to think about youngsters in terms of the grades they earn. We stick on firm labels—A, B, C, etc.—which follow children from class to class and year to year. For instance, if someone asks you what you were like in college, you might easily find yourself answering, "Oh, I was always a B student." But at the same time, you are clearly aware that a B grade average maintained over a four-year period in no way adequately describes the complex person you know yourself to be. You had highs and lows, marked interests and tastes; in some courses you got a "snap" B, while the same grade in other classes marked a struggle of Herculean proportions. Grades are supposed to label performance, but all too often they encourage us to label the performer as well.

Making grades more functional. Although you may have no choice about whether or not you will use grades, you do have a choice in how you will use them. Do your best to standardize the criteria according to which you assign grades within your classroom. You can begin by basing grades upon one of the other measures about to be discussed in this chapter. Make sure that the criteria you select for evaluating performance are definitive and more reflective of behavior changes. Select criteria which have a clear meaning for yourself and the child. If, for instance, you decide to base grades upon percentage scores, try to keep the standards for each grade consistent from one day's performance to the next; if you choose rate, be careful to specify for yourself and the youngsters in your classroom what constitutes "A" performance (i.e., 100 words per minute with two or fewer errors in oral reading) and stick with your criteria long enough for each child to know what to expect. In the process of examining the grading criteria you establish, attempt to answer the following questions:

— Is simple quantity what you're after—the child with the most volume (in class assignments; extra book reports; homework)—gets the best grade?
— Is accuracy—the number of correct responses in proportion to total work assigned (completed)—an issue?
— Is time a factor—how long it takes a child to complete a task?
— Is quality important, and how have you defined that quality?

MEASURES FOCUSING ON AMOUNT AND ACCURACY

Amount and *accuracy* are aspects of performance that teachers frequently focus upon and work to improve. The measures examined in this section concentrate directly on these two performance dimensions.

Raw Score

Definition. Raw score is simply a straight count: the actual score itself (the total times a behavior occurred) which is expressed numerically and unadjusted.

Calculating raw score. Determining raw score requires no fancy calculations. Simply count the responses that occur and total them. If you are concerned with accuracy, note both correct and error counts for a more complete record.

Arrangements for data collection. No special arrangements are necessary. If you are keeping a count on a particular behavior (e.g., "out-of-seats") throughout the day, you may find a tally sheet or mechanical counting device helpful. Otherwise, paper and pencil are sufficient.

Special focus. Since raw score is simply a record of a behavior count or performance score, it focuses totally upon quantity. Its simplicity is its main advantage. The measure demands no calculation and no translation, but provides a direct and immediate picture of performance quantity. You never lose sight of the original data, as can happen with other measures which will be discussed later.

Limitations. Because it maintains a specific focus, raw score, like every other measure, imposes certain limitations on the way in which we view performance.

— *What you see is what you get.* Unless you are extremely careful in the way in which you go about setting up assignments and in your reporting procedures, you may have considerable trouble evaluating and interpreting the raw scores you collect. Some teachers record only correct responses, finding themselves in the position of Meda Thompson, who reaches the end of the grading period to confront this score line-up in her record book:

Name	M	T	W	Th	F	M	T	W	Th	F	M	T	W	Th	F
Ralph	2	90	24	55	70	15	10	85	36	50	45	20	10	5	30

Meda is unable to determine from the scores reported what might account for incredible performance fluctuations from one day to the next. Maybe this student is extremely erratic, and plunges from heights of 90 correct on one day, to the depths of a mere 24 the next, only to soar again the following day. Perhaps the student's performance is extremely consistent. He works with nearly 100 percent accuracy each day, and the fluctuations in scores reflect instead Ms. Thompson's teaching: She assigns a quiz with only two problems on one day, 90 problems on the next, and 30 on the last. From the scores alone, there is no way of accurately assessing performance differences on a daily basis, or of weighting performance on one day compared to the next.

This problem may be compounded by summary procedures at the end of the work term. At this time, some teachers simply add and average correct scores, and then compare results to a class curve before assigning grades. The procedure is simple enough, but also distorts the picture of performance. The child who is capable of scoring high might work only on days on which a large number of problems are assigned, "beefing up" his overall average, and do little or nothing on days when only a few problems are assigned. Even though this child's performance is highly erratic, he ends up looking much better than his neighbor, who plugs along from day to day with more consistent, but never dazzling, performance.

Easy, you say. You record *both* correct and error, so there's no problem. Consider the entries in Meda's record book when both correct and error raw scores are entered:

Name	M	T	W	Th	F	M	T	W	Th	F	M	T	W	Th	F
Meda	2	90	24	55	70	5	10	85	36	50	45	20	10	5	30
	10	50	23	16	25	3	0	6	42	2	4	2	3	4	5

Chances are, you may be no better able to interpret these scores than when correct alone were given. Where you had too little information before, now you have too much. Although the information is complete, it is still extremely difficult to decipher. With effort, you may be able to see what pattern, if any, exists in this performance. But interpretation at a glance is impossible; and, even if your math skills are excellent, you may still have problems coming to any quick conclusions about this score line-up. And these are the scores of only one pupil: imagine the headaches when you multiply them by 25 or 30!

— *Communication limitations.* The same difficulties experienced by the individual teacher in interpreting raw scores can complicate communicating performance results to others. A teacher may find himself reporting to a concerned parent, "Well, Mrs. Doolittle, Randall got *three* here, *ten* here, *five* here, and *six* here," only to have the parent counter, "But, what does that mean?"

In order to make communication of raw scores easier, teachers may want to set up clear guidelines detailing exactly how scores may be interpreted and how, in turn, they will be translated into grades or other final report notations. At the same time, some teachers may decide to translate raw scores into another form, such as percent.

Making the most of the measure. Regardless of the behavior measure you choose, you want to be certain that changes in scores and response patterns are truly a function of changes in performance capabilities, rather than merely reflective of the way you have set up assignments and opportunities to perform.

— *Standard number of problems.* One way to overcome the problems of interpretation and communication just described is to standardize the number of problems you assign each day. Of course, if you give 50 chances to perform one day, three the next, and 35 on the following, it will be more difficult to determine what of a child's increased, decreased, or inconsistent response is related to his comprehension of the instructional material, and what of it simply is a response to changing task requirements. If, on the other hand, you keep problem numbers the same from day to day, your data will be much easier to interpret. Many teachers already give standard numbers of problems (e.g., 25) on daily worksheets, without, perhaps, realizing the benefits in terms of interpreting performance results.

— *Standard amount of work or performance time.* Another solution is to standardize the time a child is allowed to perform, letting the youngster complete as many problems as he can in the time allotted. For instance, you might always provide exercise or probe sheets with 50 or more math problems (more than you feel the child is likely to be able to do) and keep timings consistent (e.g., one, five, or ten minutes) each day. As the child's raw score increases, you can easily see that performance is improving. Counting a social behavior, such as "makes positive comments to peers," you would likewise make sure that the amount of time (e.g., 20 minutes) over which you counted the response remained the same from one day to the next.

Percent

Definition. Percent statements express a proportion. When percent is used to describe performance, the statement is often made in terms of the proportion of correct responses (or answers) relative to the total number of responses completed.

Calculating percent. To determine percent, set up a proportion statement like this:

$$\frac{number\ of\ correct\ (or\ error)\ responses}{number\ of\ total\ responses}$$

Then do the division necessary. For example, if your pupil Andrew completes 8 out of 9 arithmetic facts correctly, his percent correct is calculated in this way:

$$\frac{8\ (correct\ responses)}{9\ (total\ completed)} = .89$$

Decimal fractions are inconvenient and troublesome for many, so results are usually converted to a whole number statement by multiplying them by 100.

$$.89\ x\ 100 = 89\ percent$$

Data collection arrangements. As with raw score, data collection arrangements for percent are simple. To figure percent on written assignments, a pencil or pen is all the equipment necessary. If you're not good at long division, you may want to use a pocket calculator, or set up assignments to contain round numbers (e.g., problems in sets of 10s) to make the division process easier. If you are collecting data on a behavior that occurs randomly throughout a specified time period (e.g., what percentage of a pupil's contacts with his peers are initiated by the pupil himself), you may want to devise a simple tally sheet.

Special focus. Percent's special focus is upon *accuracy*, which is a crucial performance factor in most academic areas. It provides a statement of the degree of correctness, or freedom from error, with desired behavior near the 100% end of the spectrum. Percent also portrays a clear ratio between correct and error. If you know, for instance, that Ralph scored 80 percent correct on a spelling test, simple subtraction (100% − 80% = 20%) informs you immediately that his error percentage is 20. Comparing the two percentages by putting them into a proportion statement (80% : 20% = 4 : 1), it is apparent that Ralph gets four times as many right answers as he does wrong.

Like grades, percent is a measure easily communicated because it is so commonly used as a measure in our society. Everyone knows what figures such as 50 percent and 100 percent mean. The measure relates all results to a standard—100 percent—which is accepted and understood.

Limitations. Comparing performance to a "perfect" 100 provides an easily understood and communicated measure of behavior. However, the way in which percent is calculated may shape our perspective on performance in ways that are not immediately apparent.

— *A 100 percent performance "ceiling."* All percent scores are compared to a maximum performance—sometimes called a performance "ceiling"—of 100 percent. The figure implies that the child has done the best possible, and that there is no room for improvement. In terms of accuracy, as a specific assignment is set up, this statement is valid. But sometimes the 100 percent figure does not tell the whole story.

Many teachers have had the experience of working with the child who achieves 100 percent on all the assignments and tests, but he seems bored and listless. Something's obviously wrong. The record book says the youngster is "on top"; but a sensitive teacher recognizes that, in this case, 100 percent is not enough. The child is doing the best he can under the assignment limitations imposed, but he needs more. Maybe more problems, maybe harder problems, are in order. While 100 percent has a universal meaning, that meaning sometimes obscures individual performance characteristics and needs.

The same type of difficulty may be seen with a child who works at a very different level on the performance spectrum. Take the example of a severely handicapped youngster who cannot successfully lift a spoon filled with food to her mouth even once when she enters the educational program. Her teacher, not wishing to discourage the child, begins by attempting to get the youngster to raise the spoon to her lips twice during each feeding session. When she is finally able to perform correctly twice, she reaches the 100 percent accuracy mark. It's a big day for teacher and pupil alike, and a dramatic accomplishment for a youngster who had absolutely no self-feeding capabilities before. But in this case, as in the last, 100 percent is not nearly enough. The child needs to be able to perform, two, five, perhaps 20 times better than she is presently doing to even approach what might be considered independent feeding skills.

Remember, when you use percent, that the upper limit is always in relation to performance limits you have set. As a result of the way this measure is figured, performance cannot be open-ended; and whether you set the limits at two problems or 50, you must be sure that the individual youngster is being given enough opportunities to perform. Ask yourself as you set up assignments, "If the child gets 100 percent correct, am I certain that he is doing the best he can do in relation to *himself*, as well as to the performance criteria I have established here? Is he being sufficiently challenged? Is he getting sufficient practice?" Further, how does the performance limit you set apply to skill transfer and generalization? Will the number of opportunities allowed give the child sufficient encounters with the material to enable him to transfer and generalize skills to problems outside this particular instructional situation? If the answer is "no" to any of these questions, you may want to consider using a different measure entirely, or consider ways of setting the task up differently.

— *Interpreting the increments involved in scoring.* The differences that percent scores reflect from day to day are easily interpreted only if the number of items assessed remains constant from day to day. Here's why: Depending upon the number of problems given, the actual scores a child can achieve between zero and 100 percent will vary considerably.

Suppose you give a quiz containing ten items. Each item on the quiz is worth 1/10th of the total, or 10 percent. The scores a child could receive on such a test would range, then, from zero to 100 percent in increments of 10. If you deliver a similar quiz the next day, but this time offer 20, rather than ten test items, the possible scores will look much different. Each item is now worth 1/20th of the total, or 5 percent. Scores can be zero, 5 percent, 10 percent, and so on, up to 100 percent. Each increment is 5 percent rather than 10 percent. The last day of the week, perhaps you decide to make it easy on the kids, since it's Friday, and give a pop-quiz with only two problems. But look what happens to the scores: Because there are only two items, each is now worth 50 percent.

This increment difference is important whenever you compare either correct or error percentages from day to day. While one error would have resulted in an error score of 10 percent on the first quiz of the week, the same number of errors—just one—shows up as 5 percent on the second, and 50 percent on the last! Once you have recorded the percentages, unless you keep the raw score records, or a work file for each youngster, there is no way of telling whether 50 percent represents one out of two problems incorrect, or 10 out of 20, or 30 out of 60. Sometimes these differences are critical to successful instructional changes.

Making the most of the measure. Standardizing certain performance conditions will to some extent guarantee that score changes from day to day relate to changing performance capabilities.

— *Ensure a consistent number of response opportunities.* Percent scores should reflect the number of items in a task. If you consistently present 100 math problems, for example,

50 percent will always relate to 100 problems, rather than sometimes to 50, sometimes to 100. Keeping the number of response opportunities consistent over periods of time will eliminate· many of the interpretation difficulties related to scoring increments.

Trials-to-Criterion

Definition. Trials to criterion is a measurement set-up which involves establishing a criterion and then allowing opportunities, in terms of "trials," for the specified behavior to occur until criterion is met.

Calculating trials to criterion. Three steps are involved in setting up and calculating trials-to-criterion.

— Set a clear criterion, using the measure which best describes the performance you desire. (This criterion is most often defined in terms of raw score, or sometimes percent.)

— Give the child trials, or opportunities to perform. A trial may involve going through a partial or whole sequence of behaviors leading to the pinpointed response. For example, in teaching a child how to make a bed, the entire chain of steps leading up to the final behavior may be counted as one trial, with the trial completed to criterion when the bed is "made." Or a smaller segment of the total sequence may be used for trial purposes (e.g., the first step in the sequence, placing the bottom sheet on the mattress, may be considered a trial).

— Generally, a fixed number of trials is given each day, with criterion performance established in relation to the entire series of trials. For instance, ten trials are given each day; criterion is met when the child achieves eight out of ten correct. Sometimes criteria may include stipulations regarding consecutive corrects (e.g., the child must get eight out of ten consecutive responses right, or eight out of ten right for two consecutive days).

Data collection arrangements. Trials-to-criterion involves a very simple measurement set-up, requiring no more than pencil and paper as equipment. Data may be kept on a tally sheet of your choice, or you may decide to design a data collection sheet with boxes corresponding to the number of trials looked at each day (e.g., 10). Each time the child completes a trial successfully, a check is made in the appropriate box (errors need not be counted). Such a data sheet may include enough space to record data over one or more weeks' time.

Special focus. Trials-to-criterion is a popular measure for tasks composed of a series of very similar, or repetitive components (e.g., buttoning, sorting, articulation practice), as well as being a favorite measurement set-up for examining behavior chunks in areas such as self-help and prevocational training (e.g., making beds, making sandwiches, washing hair, shoe tying, setting tables).

Limitations. The set-up is based upon some assumptions which can cause problems unless they are considered carefully.

— *The number of trials is arbitrary and may be inappropriate.* The teacher is generally responsible for designating how many trials are necessary and what the criterion is. Such designations may be made totally on the basis of personal preference or beliefs about the nature of the task. Frequently, for instance, the number of trials established to examine criterion performance is a convenient quantity such as 10—related more directly to ease in recording and computation, than to the nature of the behavior observed, or the learning requirements of the youngster being taught.

— *Performance time is ignored.* Because it most frequently relies upon measures such as percent and raw score to define criterion, this procedure does not focus on the amount of time it takes a child to perform a specific task. Even though two youngsters may meet identical criteria in terms of count or accuracy (e.g., nine out of ten trials), their performance may vary widely in terms of the amount of time it takes for performance of the behavior at criterion standards (e.g., one child makes a bed successfully in five minutes while another requires 25 minutes). Time is likely to be a crucial factor in the child's ability to adapt the skill to use outside the instructional setting.

Making the most of the measurement set-up. If you decide upon trials-to-criterion, a few suggestions will help you use this measurement set-up to its best advantage.

— *Define behavior precisely.* This is nothing new. Any time you collect data, you, of course, want to be certain exactly what it is that you count as a correct response. But because trials-to-criterion often involves measurement of a chunk, and further, requires that a pupil produce several repetitions of that chunk, you need to be doubly sure that you have a complete and precise definition of the behavior. The bigger the behavior chunk you look at, the more caution must be exercised to ensure that you require the same behavior each time.

— *Identify task sequence.* Before you begin teaching a task on a trials-to-criterion basis, make sure you understand the combination of smaller behaviors this chunk contains. Sometimes performance seems to break down consistently at a certain point, but a teacher continues to require the child to attempt the entire chunk throughout endless trials. Needless to say, the criterion is never met. Task analysis will help you identify the sequence of steps involved in the total task, and help you isolate the point at which problems begin to arise. This is the new behavior toward which instruction should be directed with whatever teaching strategies you adopt.

— *Set criterion carefully.* Try to find some basis for the criterion or number of steps you require. Before you select a criterion level for an individual child, consider the nature of the task and the child involved. Make sure performance is looked at over a sufficient number of trials to guarantee that the child has mastered the skill at a level adequate for transfer and generalization to new situations and materials. (For more information on setting criterion levels, refer to earlier information about establishing performance criteria contained in Chapter 4.)

MEASURES FOCUSING ON TIME

For the measures presented in this section, *time*—either the amount of time over which behavior occurs (duration), or the amount of time it takes the child to begin responding (latency)—is more critical than any other performance dimension.

Latency

Definition. Latency represents the amount of time between presentation of a stimulus (e.g., the teacher's directions or instructions) and the initiation of pupil response.

Calculating latency. To arrive at a latency figure, you must time the interval between the designated stimulus and the pinpointed response. Suppose, for example, that you are concerned about how long it takes a pupil in your classroom to follow directions. He might do what you ask him to, but it seems to take forever to get him started. Before setting up time limits and consequences, you decide to gather a little data on the problem. For the next few days, whenever you give the child a direction, you time the interval occurring between your instruction and the child's eventual response. You start a stopwatch as soon as your direction is completed, and stop the watch immediately when the child moves to follow your direction. You keep a cumulative timing throughout the day; that is, leaving the second hand in the position at which you stopped the watch, rather than resetting it at the end of each separate timing. At the end of the day, the amount of time accumulated on the stopwatch represents a total latency figure. This figure is not, of course, very useful to you unless you can relate it to the number of directions you gave. So you also keep a tally of each time you gave a direction. Dividing the total number of seconds accumulated on your stopwatch by the total number of instructions results in the *average* number of seconds it took the child to follow a direction.

Frequently, teachers find that they need the stopwatch for other timings during the day, or they fear that the watch might be jostled and accidently reset. This makes it necessary to record each timing individually and total the seconds by hand at the end of the day. For example, you time twelve "follow directions" with this result: 10 seconds; 1 minute; 30 seconds; 2 minutes; 50 seconds; 20 seconds; 1 minute; 30 seconds; 1 minute; 10 seconds; 30 seconds; 2 minutes. To arrive at the average amount of time it takes your student to respond to direction, you divide the total seconds (600 seconds) by twelve, to obtain a resulting average or mean latency of 50 seconds.

Data collection arrangements. Since time is a crucial factor in this measure, a timing device of some sort is essential. A stopwatch will give you the precision and accuracy you need, and allow you to reset at the end of each timing or keep a cumulative timing throughout the period of interest to you.

Special focus. Latency's advantage is that it allows you to look at a dimension of performance that other measures ignore; that is the time elapsing between a cue and the pinpointed behavior. Generally, these are behaviors for which we want to see the latency figure decrease (e.g., *following directions*; *coming in from recess* or *taking one's seat* after the bell rings; *answering questions*).

Limitations. Since latency focuses totally on time, rather than on the behavior itself, we must always assume that the behavior remains consistent in nature from one occurrence to the next. This is not always the case, causing some problems unless you are aware from the beginning of the measure's limitations.

- *Doesn't concern itself with the quality of response.* A latency measure tells you only how long it takes a behavior to begin occurring. It will give you no information about the quality of the behavior itself once it starts. If you are worried about the inordinate amount of time it seems to take Stuart Little to settle down at his desk and begin working after entering the classroom, latency will answer the question of *how long*. It will not, however, tell you what kind of work Stuart does once he's "settled down." It could be he simply takes out his pencil and begins making attempts at answers, all of which are wrong. Your measure will not tell you about the quality of work Stuart produces, only how quickly he begins to produce that effort.

- *Doesn't attend to the quantity of the behavior in question.* If your problem is a child who seldom responds verbally to questions or greetings, you may want to use latency as a measure to determine how long it takes him to answer questions or return greetings directed at him. While your measure will show you whether or not you succeed in shortening the amount of time between your morning "Hello" and Shy Shelley's reply, it will not tell you *how many* things this child says in return to your greeting. Your data will not reveal the growth involved in the change from a mumbled "Hi," to "Hello! You know, Mr. Freestone, last night I saw the Sonics on TV and . . . ."

Making the most of the measure. One way to get around problems of behavior quality and quantity is, of course, to define the behavior you want to examine very carefully, and to note changes in the nature of the performance observed as you proceed. Initially, for instance, you might not really care what kind of work a child does (e.g., how accurate or in what quantity), just as long as he gets himself settled into the task. Once the "latency" problem is solved, you can begin to require more in terms of the actual behavioral content that follows the initiation of a response. You may decide that amount of time before the child responds is one problem, for which latency is an adequate measure, and that the quality and quantity of the performance initiated are another matter, best described with a different measure of behavior.

In general, the measure can be made more meaningful if you:

— *Watch the type of cue you give.* Try to be consistent. Does what you consider to be a direction remain consistent from one time to the next? Or, do you find yourself sometimes giving long, involved directions and explanations, and at others, delivering a cursory or vague directive? Do you address directions specifically to the child, or to the classroom in general; is there a difference in response time with one or the other? Constant variations in the number and type of cues you give may make data difficult to interpret.

— *What about prompts?* If the child doesn't respond in what you consider to be a reasonable amount of time, do you give additional verbal or physical assistance? Are you consistent in prompting? Do you sometimes fall into the trap of verbal prompts that might more exactly be labeled "nagging"? Do you think this makes a difference and is it accounted for in your description of the data taking set up? Remember that, when you give assistance, you are placing a limit on the time between your cue and the child's response. The resulting latency data will always reflect these limits so that the amount of time recorded is a function of your behavior as much as the child's.

— *Standardize the number of directions or cues you give.* Response variations are often a result of the number of opportunities the child has to respond. A latency data question is no different from any other in this respect. Each direction, explanation, or cue you give is another opportunity for the child to provide a correct or incorrect response. If you deliver hundreds of cues one day, and only three or four the next, you may see dramatic changes in your data which will be difficult to account for unless you are aware of this factor. By keeping the number of directions you give fairly consistent from one day to the next, you eliminate many problems of interpretation.

Duration

Definition Duration is another measure whose crucial dimension concerns not the nature of the behavior itself, but *time*. Duration represents the amount of time over which a pinpointed behavior lasts.

Calculating duration. Duration is a fairly simple measure to calculate, providing you have carefully defined the behavior you are examining. Begin timing when you observe the pinpointed behavior to start, and stop timing as soon as the behavior ceases. Your total represents the duration (total amount of time) spent engaging in the specified behavior. Duration may be summarized and documented in a variety of ways:

— *Cumulative time.* Keep a running cumulative time or take discrete time samples whenever you see the behavior occur. Add these consecutive discrete timings together, to find the total of minutes or seconds spent engaged in the behavior.

— *Average time.* Sometimes you want to determine the average length of time spent engaged in a behavior—such as average seizure episode in the case of a child who has a history of petit mal seizures. Divide the total number of minutes (or seconds) during which seizures were observed, by the number of episodes counted.

— *Percent-of-time.* Divide the total time spent engaging in the behavior by the total amount of time over which data were taken.

Here's an example of how duration might be applied. Although it occurs outside the classroom, it's one most of us would recognize. Franklin Furlong spends four hours nightly in front of his television set, watching his favorite sports and wildlife shows. His wife, Felicia, is frantic. She wants to see her game shows every once in awhile, and, most frustrating of all, is her observation that Franklin spends a large proportion of his show-time sleeping. Felicia gets fed up, and decides to show Franklin how he uses this part of the evening. She times the number of minutes her husband spends sleeping, then determines what proportion of total time in front of the television this is:

$$\frac{\textit{Minutes sleeping}}{\substack{\textit{Minutes in front} \\ \textit{of TV}}} = \frac{160}{240} = .67 \,(approximately) = 67\%$$

It's easy to see that Franklin is asleep for about two-thirds of the time he claims to be "watching" his favorite shows. Felicia convinces Franklin to listen to reason. He now sleeps through four games shows nightly.

Special focus. Any behavior whose most important dimension is the amount of time spent engaging in it can be described using a duration measure. Teachers commonly use duration to determine how much time a youngster spends in activities such as tantrumming, out-of-seat, or desired behaviors such as staying on task. Duration has also been used to examine such important problems as length of seizure episodes (sometimes in conjunction with medication studies) and is applicable to any question of strength and/or endurance.

Limitations. Duration data, like those for latency, must always be interpreted in light of the fact that this measure focuses primarily upon time, rather than on other characteristics of the behavior itself. This raises some problems.

— *Limited attention to performance quantity and quality.* Like latency, duration focuses primarily upon issues of performance time, and therefore says very little about such performance characteristics as behavior quantity or quality. Recorded duration may change, decreasing or increasing, without necessarily reflecting desired changes in behavior. Ms. Lynch, a practice teacher in a resource classroom, decided to increase a child's "on-task" behavior by giving him points on five-minute intervals if he was "working" on the task assigned. Sure enough, each time she checked, the youngster was bent over his desk, industriously putting pencil to paper. By the end of the week, Ms. Lynch was delighted to have increased the amount of time the child spent on task from five to almost twenty-five consecutive minutes; and the youngster was happily spending his earned points on free time activities. According to the measure she was using, this teacher's program was a raging success. What Ms. Lynch had not noticed, was that the child was doing less and less on each assignment. By the end of the week, he did fewer than half his accustomed number of problems; and most of those he completed were wrong. To her chagrin, the

the teacher had shaped all the accompanying "paraphernalia" of what she defined as "on-task" behavior, but none of the hoped-for results.

Ms. Lynch might have had more success, had she managed the situation differently from the beginning; for instance, always checking to make sure the child was providing correct academic responses as she was giving points. All the same, the measure she chose was designed to reflect only time "on-task"—not the accuracy or quantity of the actual task itself. Duration data by themselves would never have indicated the problems Ms. Lynch eventually had; and, depending upon these data alone, she would not have realized that a change was needed.

Making the most of the measure. In order to assure that changes seen in your data are related to desired changes in the behavior itself, define the behavior clearly and carefully at the outset. You can improve the ease with which you interpret duration data by standardizing the data collection procedures in these ways:

— *Identify behavior onset and endpoint clearly.* Even though *time* is duration's major focus, a clear definition of behavior is still extremely important. Because duration is a measure of *how long* behavior lasts, collecting accurate data depends upon knowing exactly when behavior starts and when it ends. This can be especially tricky in the case of behavior chunks—those bulky creatures who sometimes defy even our best attempts at pinpointing them precisely. A chunk may contain so many smaller pieces of behavior that unless it is defined with great care, counting and timing will be difficult. Always decide upon your behavior's perimeters—what constitutes onset and what signals termination—*before* you take stopwatch and counter in hand.

— *Keep the time period during which you observe standard from day to day.* Otherwise you run into difficulties trying to interpret data in which a child spends five out of ten minutes in his seat one day, the next day, 20 out of 100. Such data are often expressed in percent, compounding the confusion. Try, instead, deciding ahead of time upon a period of time that seems appropriate to studying the behavior, and reasonable in terms of what you can keep data on; take your duration data using the same "base" time period each day.

— *Keep the number of responses you look at uniform.* Another option is to look at only a certain number of responses each day. Decide, for instance, that you will take duration data on a limit of ten episodes of the pinpointed behavior per day. The resulting data will show that, while today ten responses took 20 minutes, yesterday, the same number took 35 minutes.

A MEASURE INCORPORATING AMOUNT AND TIME

The final measure discussed in this chapter—rate—incorporates both *amount* and *time* dimensions in its look at performance.

Rate

Definition. Rate, or frequency, is the average number of times a pinpointed response occurs per unit of observation or instruction time. For classroom purposes, rate is usually defined as number of responses per minute.

Calculating rate. To arrive at a rate figure, divide the number of responses counted by the total number of minutes during which behavior has been observed. For example, Mr. Hubbard is counting Fast Mouth Freddie's "talk-outs" during a 30-minute study period in each day's world history class. On a particular Wednesday he counts 60 outbursts. At the end of the period he figures and records Freddie's "talk-out" rate:

$$\frac{60 \ (talk\text{-}outs)}{30 \ (minutes \ observation \ time)} = 2$$

Freddie's rate for the period observed that day is two, or approximately two talk-outs every minute. Even if he observes for exactly the same amount of time each day, Mr. Hubbard always calculates the behavior frequency, rather than recording a straight count (raw score). This allows him to compare his data with similar data collected by other teachers who may not watch the behavior for the same amount of time.

To describe academic performance, both correct and error counts are frequently taken. Figuring correct and error rates gives a more detailed picture of academic performance than determining correct alone. Thus, Ms. Wimpole tallies 100 correct and 20 error responses on Frieda's home economics quiz and divides them by the 20 minutes allowed for the test. Frieda's rates are five correct per minute, and one error per minute.

Be sure to use only the number of minutes actually spent counting the behavior as you calculate rate. If behavior is counted over an entire school day, 6 hours long, the time will be 6 x 60 = 360 minutes. If for any reason you do not observe or count during a portion of that time—when the child is out to recess or in the lunchroom—be sure to subtract the number of minutes during which no observation takes place. Maybe he is still performing the pinpointed behavior; but you aren't there to see it; and keeping the extra time in your calculation procedures will skew your data.

Data collection arrangements. Because rate incorporates behavior and time, arrangements for both counting and timing of behavior are necessary. Any one or a combination of the counting and timing devices described in the following chapter may make data collection easier. The timing period is generally specified in terms of minutes, because minutes seem more functional for teachers (who frequently talk about a 20-minute work period, 15-minute recess, 30-minute lunch break) than other time units (e.g., seconds or days). For this reason, make sure that whatever timing device you choose registers minutes (or, as does the stopwatch, both minutes and seconds). The upper count limit of the behavior you are observing will determine what type of counter you select.

Special focus on behavior. Rate is a familiar and natural measure in many areas important to us. In medicine, heart rate and blood pressure are common rate measures. In economics and social sciences we talk about the rate of inflation, interest rates, population growth rate, and many other issues for which a rate measure is crucial to our understanding. But this measure is a relative newcomer to the classroom. In recent years, rate has also been demonstrated to be a practical and sensitive measure of pupil performance, with application across a wide range of academic and social behavior contexts. A number of educators have developed well-documented procedures for collecting, displaying, analyzing, and using rate data. But despite these discoveries, teachers still, on the whole, view rate as a foreigner in the classroom. In order to make the measure seem less alien, we have devoted somewhat more attention to it in this "special focus" section, than to previous, more familiar measures.

— *Rate allows examination of a wide variety of performance characteristics.* Unlike other measures, which focus primarily upon a specific aspect of performance, rate is sensitive to a variety of performance dimensions and characteristics.

1. *Speed.* The first of these is, by definition, *speed.* Many teachers say, "But I don't

really care about how fast a child works"; and yet, speed is an important underlying dimension of performance. Consider, for instance, the work of two youngsters in your classroom, both of whom are getting 100 percent correct on daily assignments. But while Terry finishes his work just under the bell, requiring 30 to 40 minutes of work time, Marie completes her work in less than 15 minutes and may spend the rest of the period daydreaming. Speed, or rate, is certainly an issue in assessing the programming needs of these two pupils.

Terry is working with 100 percent accuracy; but, in terms of speed, he appears to be challenged at his own level: You probably would not rush to give him new material. Marie, on the other hand, while demonstrating exactly the same raw score or percentage figure, appears ready for more difficult and more challenging material.

2. *Quantity.* A measure of the total amount completed—raw score—is easily derived at any time simply by reversing the rate calculation procedure. Multiply the rate figure (e.g., two per minute) by the number of minutes timed (e.g., 10) to determine the total number of responses (2 x 10 = 20) completed. If you think you will be interested in retrieving the raw score at a later date, remember that you will have to keep track of the amount of time over which each sample was taken. Standardizing time samples (e.g., all one-minute or all ten-minute) makes such conversions much simpler.

3. *Accuracy.* Rate makes it possible to describe performance accuracy. Percent of correct and error responses can be figured easily by dividing either figure by the combined sum of correct and error responses. In addition, rate can enlarge the way we look at the accuracy issue. Think in terms of accuracy pairs. While, with percent, we always assume that increasing correct responses mean decreasing errors, different error patterns become apparent with rate (e.g., correct responses go up and errors go up also, or corrects go up but errors remain the same). Such patterns allow the teacher a more sensitive look at performance, and thus more flexibility in making teaching decisions. Finally, if you wish to describe accuracy in a very simple way, merely compare error to correct rate. If, for example, correct responses are at 80 per minute, and errors at two, correct and error rates form an accuracy ratio of 40-to-1.

4. *Proficiency.* Performance proficiency is a relatively new idea whose importance has yet to be fully explored and appreciated by educators. Proficiency, or fluency, is a statement about performance which incorporates information both about a pupil's *speed* and *accuracy*. We may discover that a child demonstrates speed in his work, or that he demonstrates accuracy; but he may not be truly proficient in that skill until he is both fast and precise in attacking problems.

Simply knowing how to add accurately, for instance, does not mean that a child has developed a useful skill. Cynthia, who can solve basic addition facts with perfect accuracy at a rate of two per minute, "knows" how to add. But, she pays a painful price for accuracy. Her rate of two per minute is so slow that it is doubtful she will ever have time to demonstrate her skills in most situations calling for addition. Current data show that children need to perform basic addition facts at a rate of 30 per minute with two or fewer errors (remember, proficiency combines speed and accuracy) in order to be proficient. If Cynthia is to work in the next unit of her arithmetic text with any degree of success, or if she is expected to use these addition skills in a new or more complex context (e.g., making change) she needs to become proficient in her basic math facts. If she cannot perform this basic skill both quickly and easily, Cynthia will probably never use these skills outside this controlled classroom situation. She certainly is not likely to find math enjoyable or exciting under the present circumstances.

Proficiency is a relative term. What is proficient performance on the part of one individual may be too little or too much for another. A newscaster reads copy aloud at approximately 100 words per minute, with two, usually fewer, errors. This rate gives us some indication of an appropriate criterion for fluency or proficiency if we are considering an adult reader with years of practice. While you would not expect a beginning reader to match this level of fluency, 100 per minute with two or fewer errors is still a valuable clue as to the "desired rate" you might eventually aim for with your own readers, using appropriate level materials.

Proficiency differs not only from one individual to another, but from one skill to another. Much exploration of the issue is necessary before any firm statements about what constitutes proficiency can be made. Even so, the ability to combine a picture of both speed and accuracy in one measure is still valuable in assessing what a child has achieved.

5. *Endurance.* While speed and accuracy are significant indicators of successful performance, it is also important to know whether an individual can maintain the strength and perseverance of his performance over lengthening periods of time. Suppose a youngster is referred to your resource classroom for problems in reading. As a part of the program you establish for this child, you decide to have him read aloud to you for one minute each day, while you track the progress of his correct and error rates from one day to the next. The child's rates show nice gains. Then, one day after reorganizing your schedule, you find you can spend more time with this youngster, and you begin listening to him five minutes daily. A distressing phenomenon appears in the data: Although you are spending more time with him, the child's performance seems to be getting worse. His correct rates are decelerating, while at the same time, errors increase. What's wrong?

One definite change you can identify is the amount of time you have been asking this youngster to read aloud. You try the one-minute periods again and notice that his rates immediately "bounce back." Given the five-minute periods, there is no question that he shows more limited success. This child demonstrates a definite *endurance* problem. He does all right in the short sample, but falls down in the long haul. An educational decision you might make based on these data is that for this child to succeed in the regular classroom, he must definitely build his performance endurance; this means, to read for longer periods without significant drop in correct rates and climb in errors.

Because rate takes time into account, it allows us to contrast a child's performance on samples of a wide variety of lengths: one-minute, 10-minute, 20-minute, and even longer. We can see where the child's endurance might become a problem, and over what sample length the child is able, conversely, to demonstrate maximum capability.

An understanding of the part endurance plays has made a significant change in the behavior sampling procedures of teachers who practice extensive use of rate in the classroom. These teachers find that one-minute samples are often optimum in the collection of certain academic data, since they require minimal time expenditure on the part of the teacher, and, at the same time allow a picture of optimal pupil performance, free of the fatigue factor. This is a look at the student "at his best."

Endurance is a performance factor that none of the other measures described in this chapter account for, but it is, nonetheless, crucial to success in all areas of life. We all know that the longer we persist at a task, after a certain point, our strength begins to diminish. This is the problem with which every long distance runner must contend, and there are certainly other situations, inside the classroom and out, in which running the distance will be important.

— *Looking at performance growth.* The more performance characteristics a measure takes into account, the better able we are to observe subtle performance changes in relation to many instructional variables (e.g., introduction of new materials, teaching strategies, or even different seating arrangements). Incorporating this detailed information makes teaching decisions more sensitive to individual pupil performance.

— *A "standard" measure of performance.* Teachers who use rate commonly refer to it as a "universal datum." By that, they mean that rate can be used to examine and compare vastly different areas, whether social or academic. Other measures might tend to limit the way in which we view behavior, restricting us in the number of performance characteristics we can describe, or in the type of behavioral issues to which they are best suited. Rate, however, allows us to combine a wide range of pieces of behavioral information into one simple type of datum, or one mark on a chart, that is sensitive to a variety of different performance dimensions across a variety of different performance areas.

The fact that this measure incorporates time rather than choosing to ignore it, is a distinct advantage. *Time is a great leveler.* While this statement is fraught with all kinds of physical and metaphysical implications in our lives, for the issue of rate, it holds a very practical and simple truth. Time—specifically, the time quantity of "per minute"—is used in the rate formula in much the same way that the numerical quantity 100 is used in figuring percent: it gives us a uniform shape (behavior[s] per minute) for data from one instance to the next. This offers us the opportunity to compare the occurrences of dissimilar behaviors such as writing answers in geography, talking out, hitting or tipping desks that might occur within the same period of time. It also enables us to compare and contrast their occurrences with behaviors taking place at several other times during the day, such as in reading period or recess. Not only can we compare behaviors occurring at different times throughout the day, but we can also compare behaviors that occur over periods of dissimilar length. Rate allows us then, to make comparisons across numerous and diverse situations.

It is a great help to be able to use the same measure for all or most of the behaviors you are looking at, as it makes any effort to interrelate results much easier. Attempts to correlate changes in one behavior in terms of changes in another need not involve long and complicated study and calculations for which the teacher lacks the time. Results are generally in fairly straightforward terms. The more data you decide to collect, the more important this will be. (How many times must you switch gears to relate results or interpret your data?) Also, for communication ease, it helps to have the same measure. (What ever measure[s] you select, try to consider how the standardization of measurement might be of help to you.)

Limitations. Sometimes issues other than a behavior's rate are overriding: We don't care to incorporate the time factor in the way that rate demands.

— *Some behaviors are more clearly an issue of latency or duration.* At times, rate simply is not an issue of much importance. The teacher who notices that a child usually engages in only one out-of-seat episode each day, but that this episode may last anywhere from a few minutes to over an hour, will not find rate a very informative measure. The question is clearly one of duration—*how long*, on the average, the behavior lasts. A few rate die-hards might wrestle with a behavioral definition of the situation until it included rate: for instance, breaking a larger chunk into several much smaller pinpoints (e.g., what goes on during the out-of-seat period—bothering peers, running, dumping objects, etc.) and determining with what frequency each of these occurs; or looking at the rate of out-of-seat in several categories, such as "lasting less than three minutes," and "lasting more than 30 minutes." But this is a very strenuous sort of balancing act which results only in a lot of extra information and sometimes misses the target. If *time* is the major question of concern—whether *time during* or *time until*—a measure other than rate should be considered.

— *Rates often come in odd figures.* Teachers who aren't familiar with rate may find comparing quantities such as .2 (per minute) and .5 (per minute) hard to get used to at first, and even stranger quantities are not only possible, but quite likely: .01 (one behavior every one hundred minutes); .05 (one behavior every 20 minutes); .2 (one behavior every five minutes). These figures are just something the rate user adjusts to with practice.

— *Rate may be difficult to communicate.* Since rate is such a new measure, without an explanation of the way in which these data are collected and calculated and without the rationale behind use of rate as a measure, teachers may find it initially difficult to get others (e.g., parents, administrators, other professionals working with the child) to understand why the measure is being applied to human behavior, and child performance in particular.

Comparing rate with three familiar measures. Many teachers hesitate to adopt rate simply because it is such an unknown in the classroom. Hoping to take some of the threat out of the measure, we have included an additional exercise in this section, comparing the use of rate and three other, more familiar, measures of behavior. The following example contrasts the picture rate gives of performance with those provided when grades, percent, or raw score are used. If the example seems somewhat biased, it is because it was developed originally for workshops set up to train teachers how to use rate in their classrooms. Bias aside, however, it is a clear demonstration of the different ways in which each measure causes teachers to view performance, and the effects these differences have on the practical programming decisions teachers make.

Each of the four third-grade teachers in Stevenson Elementary uses a different measure to evaluate the performance of their pupils in basic math skills: Mrs. Jones uses grades; Mr. Smith, percent; Ms. Anderson likes raw score; and Mr. Johnson prefers rate. Timothy is in Mr. Johnson's class; but suppose, just for the sake of comparison, that for a 12-day period, all four teachers monitored Tim's performance on math work sheets. All supervise exactly the same performance, but each records Tim's performance using his favorite data system. Figure 5:1 shows a summary of their raw data for the 12-day period. (Note that the time Tim is given to complete his work changes every three days. Days 1-3, he works for five minutes; days 4-6, for 10 minutes; days 7-9, for three minutes; and for the last three days, 10-12, for 20 minutes each day.) Following is a discussion of the different picture of performance each measure provides.

Grades. The grades picture is almost entirely flat—all B's with one A attached at the tail. Reading these grades, we would be inclined to assume that Tim's performance had gone essentially unchanged over the 12-day period, and that he had achieved little if any growth. Such a conclusion might be erroneous: Tim might, in fact, be making gains in his math performance that would eventually move him out of the B bracket, and perhaps that is what we see in the A at the end of the re-

FIGURE 5:1. *Examining Math Performance Using Four Different Measures.*

Mrs. Jones: Grades	Mr. Smith: Percent	Mrs. Anderson: Raw Score		Mr. Johnson: Rate		Work Period
		Correct	*Error*	*Correct*	*Error*	
B	91%	50	5	10	1	
B	88%	75	10	15	2	5 min.
B	87%	65	10	13	2	
B	91%	100	10	10	1	
B	88%	70	10	7	1	10 min.
B	89%	80	10	8	1	
B	95%	60	3	20	1	
B	93%	75	6	25	2	3 min.
B	90%	78	9	26	3	
B	90%	180	20	9	1	
B	88%	140	20	7	1	20 min.
A	100%	20	0	1	0	

cording period. Or maybe the A is simply a fluke, an isolated incident that will be followed again by a succession of Bs. These data simply do not give us enough information about the quantity of work completed each day, the actual quality in terms of correct and error performance, or about the comparative amount of time in which Tim finished the assignment from one day to the next, we have no way of knowing where on the continuum of "B-ness" each grade fits. We can see that Tim must perform somewhat better than the average student in the class, but how much better—or in what way better—we don't know. If he is bettering his own performance just a little each day, we don't know that either. Mrs. Jones' system gives us no indication of the basis on which she assigns a B.

Percent. Mr. Smith's percent figures are more sensitive to performance change: Some fluctuations are apparent. Far from being flat, according to a percent measure, Tim's performance demonstrates quite a bit of fluctuation, bouncing up and down in the high eighties and low to middle 90 percent range. The percent measure focuses, of course, upon the accuracy of Tim's performance. Although the teacher records only percent correct, we can easily derive the error percentage simply by subtracting the correct figure from 100. The figures seem straightforward enough on the surface; however, they can be misleading. Because Tim scores 91 percent on both the first and fourth days, we might assume that his performance on these days is equal.

Remember, however, that percent may mask certain task requirements and performance variables, such as the number of problems or items assigned, and the amount of time over which work occurred. If you compare the percentages achieved on days 1 and 4 with the raw scores for those days, you can see that task size varied enormously. Although Tim's score is 91 percent on both days, the amount of work he was required to do on day 4 is approximately double that of day 1. Examine percentages for days 2 and 5, on the other hand, and you will discover that while the *amount* of work Tim does on both days is fairly comparable, on day 5 it takes him *twice as long.*

It is interesting to note, too, that the day on which Timothy scores 100 percent—the ceiling for this measure, and supposedly representing his best day—Tim is not, in fact, doing his best. A look at the raw score shows that Tim actually did fewer problems on this assignment than on any other. At the same time, it took him *longer* to do the problems (on day 7 he scored 95 percent on an assignment requiring three times the amount of work, and performed in approximately one-seventh the time).

It may be extremely important for Tim's teacher to know what variables combine to give Tim his best chance at accuracy. Without some reference to the raw score and to the amount of time used for performance, these variables can be lost. You may or may not consider such factors important, but you should be aware that although percent provides a more sensitive measure of individual performance differences than grades, it still excludes information on certain variables that may demonstrate a marked effect on the quality of a child's work.

Raw score. Raw score has the advantage over percent of offering an easily accessible picture of total amount, since it is nothing but a notation of amount completed (in this case, both correct and errors). Raw score lets us see clearly and directly the daily fluctuations in the number of problems Timothy was assigned and to analyse for ourselves whether or not assignment size influences his correct performance. We can, if we wish, compute a percent statement, as well as the ratio of corrects to errors from the raw figures provided and thus have a measure of both quantity and quality.

At first glance, however—and this may be all you have time for with 30 youngsters in each class—you may find it difficult to see performance trends or make comparisons. This type of data system can be unmanageable, and it is easy to see why teachers often opt for a cleaner system, such as percent or grades. As with percent and grades, the time dimension is excluded when raw score is the measure.

Rate. The rate measure, although an unfamiliar one to many teachers, offers many advantages in this situation. It retains information about performance dimensions such as both quality and quantity.

- *Quantity* of correct and error can be easily computed simply by multiplying the rate correct or error times the number of minutes in the work period. On the first data day, for instance, Tim works at a rate of 10 per minute correct for a total of five minutes. Multiplying the rate by the time (5 x 10) gives us 50, the number correct under raw score.
- *Quality*, or accuracy, is quickly derived by adding correct and error rate and dividing the total into the correct rate. On the first day, Tim does a total of 11 problems per minute (10 correct plus one error); dividing 11 into 10, results in a correct percentage of 91—the same figure obtained by doing a percent operation on the raw scores. The correct and error pairs are easier to read and compare. The ratio of correct to error is easily determined, simply by dividing correct by error rate. On the first day, Tim has correct and error rates of 10 and one respectively, making the ratio of correct responses to errors 10:1.

In addition to information about quality and quantity, the rate measure lets us look at one more dimension of Tim's performance: *speed*. We can see that in certain of the three day stretches (e.g., days 7-10), speed improves, while in others (e.g., days 10-12), speed decreases.

Just two simple figures—the correct and error pair—offer a combined picture of quality, quantity and speed. We call this combination a measure of *proficiency*. If we look carefully at correct and error pairs during those days when the correct rate increases from one day to the next, we can see that correct increase is often accompanied by error increase. While Tim is getting faster, he is not necessarily getting more proficient.

By comparing rates and the length of the work period each week, it becomes obvious that *endurance* is another important factor in Tim's performance. That is, his rates are consistently higher when the working period is shorter. In fact, Tim shows a general increase in correct rates (growth) during the days in which work periods are shorter, and a decrease in correct rates from one day to the next during the three day stretches with work periods of 10 and 20 minutes. If we can believe that the type of problem remains consistent from day to day, knowing something about the length of the work period gives us a very important clue about Tim's skills.

This measurement comparison exercise points out more than anything else, how important it is to set tasks up with a maximum of consistency to get the most meaningful information out of whatever data system you select. Many of the limitations on data comparison and interpretation posed here could have been avoided by some attempt to standardize the measurement base. That is, if either number of problems, or amount of time were held constant from day to day, we could make much better data comparisons. If, for instance, Tim's assignment always included 120 problems, or if it always included only 20, the fluctuations seen in performance would more likely be related to changes in Tim's ability to solve arithmetic problems, than to changes in the assignment size.

Making the most of the measure. Rate, like each of the measures described, improves in the sensitivity with which it describes pupil performance changes when care is taken to standardize data collection.

— *Fixed number of problems.* One way to standardize is to present a specified, fixed number of problems or opportunities to respond, recording the amount of time it takes the child to complete all items. Although this means of collecting rate data is sometimes used in the classroom, it is not a preferred one, for the same reasons outlined in the discussion of "locking" performance in the *probes* section of Chapter 3.

— *Fixed amount of time.* In a second, more favored method, time, rather than opportunity to respond, is prespecified and fixed. For instance, a teacher may decide to look at a behavior such as *asking questions* over a 30-minute session each day, counting the total questions that occur within this prespecified period. Or the teacher might choose to see how many answers on a specific assignment a child could produce when he is given more opportunities (e.g., 100) than he can deal with in the time period allowed (e.g., one minute).

CONCLUSION

Rather than make a blanket recommendation about any of the measures discussed, this chapter has tried to illustrate the peculiar quirks and idiosyncrasies, as well as the advantages and disadvantages each measure holds for teachers who want to use behavior data as a basis for instructional decision making. It should be clear by now that some measures are much better suited than others to examining certain types of behavioral issues. *Duration*, for instance, might be your best bet for describing a severely handicapped youngster's increasing ability to maintain his grasp on a spoon. *Rate, raw score,* or *percent,* on the other hand, would probably do a better job of depicting how often and with what amount of accuracy a youngster is able to bring a spoon containing food to his mouth. To be able to identify the most appropriate measure is an essential step in answering any behavior question.

If you have not decided yet on the measure, or measures, you will employ in your classroom, review the choices available to you, making sure that the measure you select is the most sensitive to those dimensions of performance you consider important. Ask yourself whether or not the way in which your measurement problem is phrased gives you any clues, keeping in mind the *definition* and *special focus* of each of the measures which have been presented. For instance:

— I wish I could get the child to sit in his seat *longer* (duration).

— I wish I could get him to do these calculations more *accurately* (percent).
— I wish she were more *proficient* at working those formulas (speed + accuracy = rate).
— I wish she wouldn't take so *long between* the starting gun and hitting the water (latency).
— I wish he could type a little *faster* (rate).
— I want to know just *how many times* she's getting out of her seat (raw score).

Make certain you know how to *calculate* each measure: know what pieces of information are required to perform the calculations, and how these pieces are put together (e.g., Rate needs two pieces of information—amount and time—which are related by dividing amount by time). Review *data collection arrangements*, to decide whether or not they are practical for your situation, and what adjustments in your teaching routine will be required. Decide whether the *limitations* of the measure present any problems, and, if so, can these be overcome? And finally, determine how you can *make the most* of whatever measure you choose, standardizing instructional formats, and data collection activities wherever possible. Finally, remember that no measure is any better than the pinpoint it describes. The more carefully and precisely you define the behavior you want to measure, the more meaningful your data will be whatever the measure you use.

Chapter 6
Collecting Classroom Data

Many of the measures described in the previous chapter require certain extra steps when it comes to data collection. Latency, duration, and rate data, for instance, involve timing as well as counting behaviors. If—like most teachers—you are always busy, often short of time, and frequently short of help, you are probably asking yourself how you can be expected to count or time the responses of even one child, when you have a whole classroom to manage. As a novice to measurement, you may fear that incorporating counting and timing in your classroom, besides being extra work, will make your teaching artificial or mechanistic. Perhaps you've seen teachers in a research setting who are taking data on a large number of behaviors all day for each child they work with, and you wonder anxiously if this is going to be required of *you*. These questions and anxieties—what you might call fear of counting and timing—are natural to the beginner.

Relax. The measurement strategies introduced in this chapter are not intended to complicate an already complex classroom existence. Counting and timing should not be artificial appendages tacked on to your current teaching style. Just the opposite is the case. The measurement procedures presented here are designed to be incorporated as simple and natural components of the activities you are already familiar with.

Using these procedures does *not* mean that you will have to change the way you teach. But they can help you put your teaching into an entirely new perspective. The procedures provide a framework that allows you to sort through what you are already doing to identify your best, most effective, and most easily implemented strategies.

This chapter introduces measurement tools and techniques that are inexpensive, easy to use, often imaginative, and as appropriately and directly related to the requirements of individual learning tasks and individual learners as possible. Exposure to the wide range of options available, with a little practice in using them, can make counting and timing an exciting and valuable part of your teaching repertoire.

BEFORE YOU START TO MEASURE

Assessing Your Current Measurement Status

At this point you must decide when and where to introduce measurement procedures within your classroom routine. Before making a rapid series of changes, take time to analyze what is already going on in terms of measurement procedures in your classroom. You may be surprised to notice how much unconscious measurement activity is already occurring; and often, just looking at the status quo helps you to define more easily what needs changing or amplification.

What are you already measuring? Chances are, you engage fairly frequently in activities such as recording grades in grade books, preparing performance progress reports for parents, or summarizing material to meet school and district report requirements. In addition to what you yourself record, you may find that you have your children report results to you, or that you provide frequent feedback to them on performance quality. Can you identify the types of performance, or particular performance areas (e.g., academic, physical, or social adjustment) for which you already make arrangements to measure, record, and summarize performance characteristics or capabilities? These are "naturals": situations in which you are currently implementing some kind of counting, recording, or reporting procedures which, with just a little added work—enhancing or modifying present measurement techniques—may yield much more information.

How frequently does data collection take place? How many times during a day or in the space of a week do you find yourself collecting performance information? How long does it currently take you to gather and record this information? Do you do most of the data collection yourself; do you ask the youngsters in your classroom to provide some of this information to you; and how accomplished are you and your students at collecting and recording the performance data you currently use?

What, if anything, would you like to change about the way you collect and use performance information? Are you as systematic or consistent as you would like to be? Would you like to give more frequent or more immediate feedback to your students about their social or academic performance? Would you like more information about a particular area or behavior to help you make better, more informed decisions? Would you like your pupils to take more responsibility in the measurement process?

What situations lend themselves most to new and additional measurement efforts? Can you identify programs, skill areas, and performance situations already in operation that seem ideally suited to adaptation or implementation of one of the measures introduced in the previous chapter? If not, what kinds of simple modifications would make introduction of measurement to those programs or areas possible?

Even if you are by now convinced beyond a doubt of the need to measure, and you are eager to make changes in your own teaching procedures that will enhance your data taking capabilities, remember that it's sometimes easy to make extravagant measurement plans, only to find yourself initially overwhelmed and eventually disenchanted. You may choose too many measurement targets to begin with, or fail to take certain management problems into account, and find that you're in water over your head, with your measurement dreams turning into a data collection nightmare.

Test the waters before you leap in. *Think big, but start slow.* Begin with just one—two at the most—problems or questions for which you think collecting some data would help you in your decision making endeavors. If you could change one thing about the performance of one child, or maybe one thing for the entire group, what would that be?

In Chapter 2 we encouraged you to establish priorities for selecting behavioral pinpoints. The

All Wrapped Up in Counting and Timing

same priorities hold as you devise strategies for measuring these behaviors. Select the behavior whose addition, elimination, or remediation will make the most impact upon your classroom or upon the performance of a particular pupil and start with that. If Randy's "talk-outs" or "out-of-seats" spell chaos for you and promise eventual placement in a special learning setting for Randy, start with these. You might try a device as simple as a tally sheet, attaching a small square of paper to Randy's desk, or to a clipboard positioned at some convenient place in the classroom. Each time Randy engages in the behavior you have specified, you make one mark on the sheet. At the end of the day, total the marks and record them on a raw data sheet. If you chart your data, they may be either directly recorded on the chart or transferred from the raw data sheet at a later time.

Decide the size of the sample: How much behavior is enough? When teachers first think about collecting behavior data, they often wonder exactly how much behavior is enough to give them the information they need for making educational decisions that are on target. They may envision themselves "glued" to a particular youngster, working frantically all day to avoid missing a single critical occurrence of a pinpointed behavior, at the expense of all the other children in the room, to say nothing of their own nerves. This kind of zeal is seldom necessary.

Unless you plan to follow a child from his first waking moments until he returns to sleep at night, all you are ever getting is a *sample* of the behavior you have pinpointed. A sample is perfectly adequate, provided that it is representative. As a teacher, it's up to you to decide what kind of behavior sampling procedure is the best, giving you a look at a particular behavior which is both accurate, and yet, involves the least amount of effort, considering the busy day you face.

Decide how many opportunities to give for the behavior to occur. Some behaviors occur more or less in a "free state." Hitting, talking out, and getting out of seat, as examples, are not behaviors for which you normally provide opportunities to occur; they happen spontaneously and randomly, and usually when you are least prepared for them. Other responses, however, need special arrangements made for them if you want to collect consistent information. Most academic behaviors fall into this category. While a child may read sporadically throughout the day under a variety of conditions (e.g., directions for math assignments, songs in his music book, signs in the lunchroom) you are likely to get the best and most consistent information about a child's reading skills by monitoring reading during a program set up to control for materials and other conditions.

For all such behaviors, you need to decide how many opportunities should be arranged for the behavior to occur (e.g., How many arithmetic problems should you give; how many geography questions should you ask?). Allowing a child to respond to 200 problems of a particular variety will give you a much different look at behavior than observing his performance on 20. The child who can handle two problems quite easily, might perform very differently if he must do 20, or 100.

Decide when and how long to sample. Even if you are using a measure such as percent or raw score, in which time is not, by itself, a performance factor you're concentrating on, time still figures importantly in data collection. Considering the issue from the standpoint of classroom management alone, it's easy to see that it would take you considerably longer to administer 200 problems than 10. Perhaps the first, and simplest, question which must be resolved concerning how long to sample is how much time you feel you can devote to looking at the quantity of behavior you have specified.

More than being simply a matter of convenience, *when you sample* and for *how long* are both questions which will play an important role in determining whether or not you get a representative sample of behavior. These questions are answered in part by looking at certain dimensions of the behavior itself: Does it occur at a high or low frequency; does the behavior occur all day long, or only at certain times; is it consistent or erratic; is it social or academic?

— If the behavior is one that seems to you, just from off-hand observation, to occur fairly consistently and frequently over the entire day, you won't need to track it from 9 to 3 each day to get an accurate picture. Instead, you might single out just one period (e.g., 15 or 20 minutes) at a consistent time each day and count each occurrence of the behavior within that period.

— Sometimes, you'll notice that a behavior seems to occur most frequently—or perhaps, only has occasion to occur—during a specific part of the day. *Throwing food*, for instance, would be most likely to be counted during a half-hour lunch, or a ten-minute snack. *Sharing toys* would most likely be observed during free time, recess, or other play periods, as would certain other social interactions. Behaviors such as *contributing to class discussions*, or *initiating questions* would be most appropriately and easily identified during group discussion sessions. When behavior is hard to pin down, you may want to spend a few days taking counts during 10 or 15 minute time blocks throughout the day, to isolate those specific times when the behaviors seem to be occurring most consistently, or to be most annoying. These preliminary data will help cue you in to the periods during which it is most important to track the behavior you have in mind.

— For behaviors that occur at a very low frequency and appear extremely erratic, you might need data collected over the entire day for an accurate picture of what's happening. For example, you might notice Light-Fingers Francine helping herself to her playmates' erasers, bubble gum, and other prized possessions. This happens only two or three times during an entire day, and there's no predicting when she'll strike. Sometimes two or three days may go by when Francine keeps her fingers entirely to herself. But the problem is persistent and troublesome. Warnings and even scoldings have had no effect, so you decide to take some data before setting up a formal program.

In a case like this, where the behavior is sporadic and infrequent throughout the day, but where every occurrence matters, you really need to count each time you see the behavior happen.

A low-rate behavior is often much more difficult to deal with than a high-rate one. You have less chance to interact with the behavior, less chance to deliver a consequence, and therefore less chance to change it. With a high rate behavior, conversely, it won't really matter if one or two happen and you don't notice. But low-frequency behaviors need a strong and immediate consequence. You have to watch them carefully and can't afford to let opportunities for interaction slip by. Although this means keeping a count over the entire day, since the behavior doesn't happen very often, recording it each time you see it won't put an excessive counting burden on you.

— Whether a behavior is social or academic may determine when, or how long, optimum data-collection periods will be, as well as the type of counting devices you choose, and other preparations you make.

Behaviors generally categorized as "social" are usually observed over periods of longer duration. Although a problem social behavior, such as swearing, may be associated with a particular activity and isolated in occurrence to a specific time period, it is likely that this behavior—as would many other social behaviors—occurs sporadically at various times throughout the child's school day. You would, therefore, count a child's "swears" throughout the entire day or for time blocks of at least fifteen to twenty minutes, perhaps several times during the course of the day.

Although social behaviors may be relatively unpredictable, academic behaviors are generally planned for: You attempt to define and control the time slot, setting, and conditions under which most academic performance occurs. For this reason, academic behaviors can usually be monitored within samples of much shorter duration.

You may wonder exactly how long an academic timing should be. For instance, if the child's math period is 50 minutes long, would you expect to figure data based upon the entire 50 minute period? The answer is, that the timing should be long enough to be functional in relation to the behavior monitored. You may deliver instruction over a period of considerable duration, but you needn't necessarily measure performance the whole time.

Your instructional objective can help you greatly in determining the length of the timing. As you consider your eventual objectives for the child, in what kinds of situations will generalization and transfer take place in? To go back to the math example just considered, the instructional period itself may be quite long—including, perhaps, a 20-minute period in which the children work at assigned problems in their text. But as the child uses his math skills outside the classroom—to compute his savings, to figure change—he will need to demonstrate skills quickly and accurately in a much shorter period of time. Thus, a very short timing may be quite adequate for a representative sample of behavior. Earlier, when we discussed probes, we spoke of samples as brief as one minute. The one-minute sample is used frequently in basic skill areas with rate as the measure. It has the advantage, in addition to being quickly and easily carried out, of getting around fatigue problems which often plague performance when longer samples are employed.

Three quick rules will help you select sampling periods that are appropriate and convenient:

— *Pick a time period related to the behavior itself.* Consider the critical dimensions of the behavior (e.g., frequency, consistency), and attempt to tie measurement decisions clearly to the behavior objectives you have defined.
— *Pick a time period that is manageable.* Only you can decide what will fit into your schedule and what won't. Try not to pick the busiest time of the day for counting and timing. (If you have no choice but to measure when your classroom looks like Grand Central Station with all the trains pulling out at once, some of the hints under data collection strategies and aides will be useful.)
— *Pick the least amount of time that will give the most information.* Don't feel that you have to watch a behavior all day in order to get useful information. Remember that short, carefully selected and arranged timings frequently offer as much or more meaningful information.

Decide whether to take a continuous sample or to use time-sampling procedures. If you take a running, or continuous sample of behavior, you will be counting the behavior every time it occurs within the time period specified (e.g., You decide to watch Andrew for a 20-minute period each morning and count each time he interrupts others within that period.). This type of sample has the advantage of providing the most accurate picture of behavior, depicting every instance of the behavior on a one-for-one basis.

Time sampling, on the other hand, involves recording at prespecified periods the occurrence or lack of a behavior. Time sampling is a procedure common to much of the most widely read behavioral research, and is found, as well, in some research classrooms set up to study behavioral issues pertaining to child development. Three distinctive types of sampling techniques are commonly used:

— *Momentary time sampling*: At the end of a specified period of time, an observer records whether or not behavior is occurring at that precise moment.
— *Whole interval*: Behavior must be occurring throughout entire interval specified in order to be counted.
— *Partial interval*: Behavior need occur only once at any time during interval in order to be

counted.

Of these techniques, the most popular is momentary sampling. As the definition says, this strategy involves cueing-in at certain intervals and noting whether the behavior is occurring at the moment of observation. Thus, the person using the procedure need look only for a very brief instant, and only at the prearranged intervals in order to take data. Momentary sampling is frequently implemented to examine self-help or vocational skills.

In practical settings, as opposed to ones in research, time sampling is often used for long tasks of a repetitive nature, and in situations where the data collector is likely to be so busy with other activities that it is difficult to see or count each individual occurrence of the pinpointed behavior. For example, staff in a prevocational restaurant training setting might use time sampling procedures to monitor dish-washing or table-setting, finding it easier to cue in and record data only at certain periods or intervals throughout the task's duration.

Social interaction behaviors, such as "cooperative play," are also a frequent subject of time sampling procedures. A teacher who is responsible for monitoring a horde of active preschoolers at recess can still manage to take data on the behaviors of a single one (or maybe more) of his or her charges using, for instance, a *momentary time-sampling* strategy.

The teacher may decide to look at Mikey's toy-sharing behavior. *Momentary time sampling* is chosen as the strategy. (Indeed, monetary sampling seems to be a most common interval choice throughout behavioral studies.) The teacher divides the recess period into five three-minute intervals. At the end of each interval, timed on a stop- or wristwatch, the teacher focuses attention on Mikey. Is he sharing a toy? If he is, then he receives a "+." If he is not, the teacher records a "−" on the record sheet. At the conclusion of the period, the teacher may record the data in terms of percent of correct (sharing) responses, or simply as a raw score (total number of *shares*).

While momentary sampling can cut down on the amount of observation time you spend, whole and partial sampling techniques demand your more consistent attention. The data-collection load is somewhat eased by the fact that with either one of these procedures you need not be recording continuously, but you have to watch for the occurrence of behavior throughout the specified period, nonetheless. (In the case of *whole interval sampling*, you must be able to confirm that behavior has occurred during the entire interval; for *partial sampling*, you can stop observing as soon as the behavior occurs; but, unless this is early in the interval, you will be occupied pretty much throughout.) With either procedure, therefore, teachers frequently find they must resort to obtaining the assistance of an independent observer who can devote attention solely to the sampling task.

At a glance, many teachers think that interval sampling is the perfect solution to data collecting problems: they need record data only at prearranged intervals which they can set in accordance with their own teaching activities. Time sampling has drawbacks, however. In addition to those just described for partial and whole sampling procedures (e.g., the necessity to keep your eyes glued on the behavior or use a hard-to-come-by assistant observer), to use time sampling you have to obtain or create a cueing device and listen or watch for the interval cue. This may be distracting to some.

In addition, you should note that each of the three time sampling procedures generates a slightly different picture of how frequently a designated behavior appears to occur within the specified observation period. In fact, the very same behavior, recorded within the same period of time, depending upon whether it is sampled according to *momentary*, *whole*, or *partial* sampling techniques may appear to have considerably discrepant levels of occurrence. Three differences have been demonstrated very clearly in applied behavior research.

— Data seem to indicate that a *momentary sampling* procedure will represent discrete occurrence of behaviors as happening sometimes more often, sometimes less often, than they actually do during the total observation time.

— If we employ a *whole interval* sampling technique, behaviors generally appear to occur less often during the period—after all, they must occur during the *entire* interval to be counted.

— A *partial interval* sampling technique usually represents behaviors as occurring *more often* than they actually do.

These discrepancies occur because the interval reflects how often you are looking, rather than how frequently the behavior is actually occurring. The length of the sampling interval obviously affects the picture you obtain. Sampling difficulties can be lessened by adjusting the length of the interval to correspond more closely with the characteristics of the behavior (i.e., how often it occurs). In general, however, teachers find that taking continuous data over an appropriate period of time yields a truer picture of performance.

DATA-COLLECTION STRATEGIES

Often measurement does not take any special arrangement. Many of the activities and exercises used on a frequent or daily basis will readily provide the data you need, by incorporating a measurement procedure within the structure that already exists (e.g., Ask children to help count up correct and errors at the end of the period; tally correct and error responses yourself during a question asking period.) The following suggestions are designed to show how you might modify the present program structure for data collection purposes.

Organizing Data-Collection Procedures

Arrange the classroom to enhance counting and timing. You may already have set your classroom up into stations or interest areas geared toward individual instructional components. Examine classroom arrangement with an eye, also, to measurement ease.

— If you are collecting data on a single child, be sure he is sitting in a strategic location, making it easy for you to see and count the pinpointed behavior as well as to give performance feedback.

— To collect data on the performance of several children working within the same instructional unit, set up data "groups" whose individuals can share counting and timing devices. Youngsters can often help each other correct, count, and record responses, as well as carry out timings, if time is an issue. Such groups work especially well, in fact, for children engaged in programs in which timed samples of performance are taken. All youngsters participating in one-minute samples, for instance, can begin and end timing simultaneously. It's not necessary that they be performing in the same part of a program, nor, for that matter, even in the same program. Each can work at his own level if materials are organized and standardized in such a way as to permit timing periods of consistent lengths across levels (e.g., All math worksheets are set up as one-minute timings).

— Take advantage of existing groups for data-collection purposes. If you have already divided your class into instructional groups for certain areas (such as reading), examine how you might use these groupings for data-collection purposes. Suppose you have your reading groups—the peacocks, sparrows, and crows—all selected. While the crows get their instruction time with you, the two other groups do independent seatwork. Consider ways by which you might use this time with the crows to examine individual performance more carefully. Is it possible to take data on something that an individual is already doing for you under such circumstances, such as reading aloud or answering comprehension questions?

— Independent seatwork time provides an excellent opportunity for data collection. To gather math data, for example, you may decide to conduct group instruction at the beginning of each math period, then devote 20 to 30 minutes to independent seatwork. While children are working at their desks, you might visit some individually, or call certain children back to a special data corner, to conduct short probes. An interaction as brief as a one-minute sample gives you a chance to monitor performance on a daily basis, and may give your youngsters a chance to ask questions about new concepts or skills with which they are having difficulty.

— You may want to set aside special data collection periods. Designate a specific period each day for data collection. During this period, encourage children to contact you independently for counting and/or timing of any of a number of skills for which performance criteria are established. The youngster who must complete a multiplication probe with 30 per minute and no more than two errors, for instance, may decide he's ready to "go for it." He can contact you at your desk, in the data corner, or ask for a contact at his own desk and request a timing. Learning that he is ready to go on to a new section may be a great way to end the day.

— Try a special measurement corner, where you keep all your counting and timing devices in one place. During a study period, individual youngsters can report to the corner for reading samples, to have seatwork corrected and recorded, etc. Even children for whom no specific behavior is being recorded can be encouraged to use the corner, to make it a special activity area youngsters enjoy.

Organize counting procedures. Some behaviors (e.g., social behaviors, such as talking out or getting out of seat) necessitate direct counting and recording procedures. That means you must be available to observe and count them whenever they occur. As you make arrangements for gathering such data, make sure you are free to count, and time, if necessary, whenever the behavior demands it.

This does not mean, of course, that you are required to stand over the child's shoulder constantly throughout the day. Wrist counters (available at most sporting goods stores) and other mechanical devices make it easy to count obvious social behaviors quickly and discretely from a considerable distance, while engaging simultaneously in the completion of other teaching tasks.

— It is not always necessary to be present at the moment a behavior occurs in order to record it. If you are measuring a "write" behavior (e.g., numbers, letters, statements), you can leave the counting and recording to any convenient time after the behavior has taken place. Such behaviors leave a "trace"; that is, the written record does not fade before you have a chance to examine total responses. "Say" responses, too, can frequently be managed so that they leave a permanent "print" for later counting. To record oral reading responses, for example, teachers frequently ask pupils to read aloud into a cassette recorder, listening to and analyzing random samples at some other, more convenient time during the day.

— Data collection can be a built-in component of many worksheets if precounted pages are used. Precounting simply involves noting ahead of time the total number of responses involved in an assignment. If you devise your own worksheets or other supplementary materials, you may find it efficient to standardize such materials, keeping the number of problems per row and the number of rows per sheet (in a math drill sheet, for instance) always the same. (Several commercial programs are also now available, which can help save considerable time here [e.g., *Supermath*, Special Child Publications, 1981].) Some teachers like the idea of printing a running total of problems or responses at the end of each row, making it easy to determine at all times exactly how many problems of the total given have been completed correctly. Leaving a specific and consistent place to record time, correct and incorrect responses, and date information makes recordkeeping faster and easier.

— Look for counting devices among the paraphernalia you may already be using. Often the stars, chips, or stickers children collect for meeting specified performance criteria make handy counting devices, while at the same time, serving powerful reinforcement value.

Organize timing procedures. A number of strategies will make timing a much easier process.

— *Encourage youngsters to conduct timings independently.* As with counting activities, you may find that timings can be carried out without your direct presence. Prerecord timings, accompanied by instructions, on a cassette tape for independent use by a single student or a group of youngsters. Sometimes less accurate, but just as easy to use, are the timing devices described in the next section, such as kitchen timers, buzzer clocks, and clock stamps.

— *Set up a timing routine.* Decide what length of time best suits the particular performance you intend to count and the measure you are using. Try to be consistent, both in when you count and how long you count, each day. Keeping timings of a consistent length is one way to standardize data, making performances more easily compared and interpreted. Further, *when* the child engages in an activity may have a significant role to play in how he performs: Make certain that you select a definite time each day during which you will examine the child and use the same period each day. Your data will be reasonably comparable and, thus, easier to interpret. Also, the timings themselves will become an automatic part of your routine.

Involving Your Youngsters in Their Own Data Collection

It's always important to encourage youngsters to be self-managing. After all, each child's successful and independent functioning is one of our major goals. The more opportunities we give children to look at themselves, to be responsible for defining and monitoring their own behavior, the easier we make it for them to reach such a goal. Involving children in data collection is one way of building and strengthening social and academic awareness and independence.

— *Teach the child to count his own work.* With the right set-up (e.g., portable timers, precounted pages, etc.) many children can learn to do their own counting and timing. Children enjoy manipulating counting and timing devices, and gain a sense of accomplishment when they are able to determine for themselves how close they come to meeting the criteria you have established for them. For some children, just being able to participate in this way can bring about an amazing change in behavior.

— *Use cross-age or peer tutors to perform simple timings and corrections.* Children can learn to help each other in data collection with minimal instruction and practice. Teaching children to count and time for themselves or for friends, as well as to monitor and count their own social behaviors, frees you for other tasks and, at the same time, involves your pupils in independent management and analysis of their own behavior.

— *Use a peer behavior game.* To take care of worrisome social behaviors that most of the youngsters in the classroom share, institute a behavior game. Divide the room into two or more teams, specifying problem behaviors for the entire group (make sure the behaviors are clearly defined and limited in number). Each time a specified behavior occurs, mark it on the board beside the team member's name. At the end of the day, the team with the least marks (for deceleration targets) or the most (for acceleration targets) wins. If you choose, winning can have a variety of immediate (or eventual, such as at the end of the week or quarter) consequences. Encouraging the youngsters themselves to come up with behaviors they want to see more or less of, in order to make the classroom a better place to live, may increase individual involvement in the game, as well as making the children more aware of the behavioral components of social success.

DATA-COLLECTION AIDS

Having assessed the measurement procedures you currently employ, and having considered the elements involved in revamping or adding to your present activities, you are ready to choose data-collection devices best suited to the data-collecting problems you confront.

Counting Devices

As you plan how to collect data on the behaviors you have pinpointed, one of your major decisions will concern what counting devices will allow you to monitor behavior most efficiently and effectively. Your ultimate choice may, in fact, greatly influence the amount of data you find yourself able to collect. A wide range of options is available, including both commercial devices and and counting tools that you design and construct for yourself.

The following are criteria which may influence the ease with which you carry out counting activities. The flexibility of a particular device in allowing you to perform other, simultaneous teaching activities is especially important.

— *Rapid action:* If you're counting high-rate behaviors or a large number of responses within one assignment, rapid action is a must.

— *Figure capability:* Will the counter you choose let you keep an accurate count without requiring that you reset the figures constantly?

— *Reset capability:* How is the reset procedure carried out? If you make an error, is it possible to reset quickly without losing the entire count?

— *Portability:* Is the device small enough to be easily carried with you from one activity to another?

— *Simplicity:* Is the device easy to use, allowing you to count and carry on other activities (such as providing verbal feedback and praise) at the same time?

— *Durability:* Will the device take the wear and tear of normal classroom use and constant resetting?

Of commercially available counting devices, the following (see Figure 6:1) meet most of the ease-of-use criteria outlined earlier.

The wrist counter. Devised originally as a golfer's counter, the wrist counter is available at sporting goods shops or in the sports section of your favorite department store. Since the counter is

FIGURE 6.1a. *Wrist Counter.*

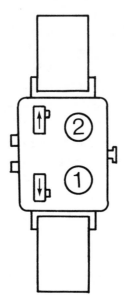

FIGURE 6.1b, *Hand Tally Counter.*

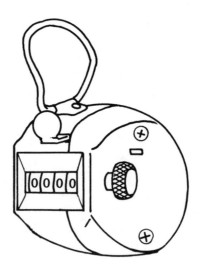

worn on a wrist band or bracelet, it is easily assimilated as a part of your everyday attire. It can be worn wherever you go throughout the teaching day. In fact, you can wear several counters at a time, if you find yourself ready to count more than one behavior using wrist counters.

Each time you see an occurrence of the behavior you have pinpointed, simply push the button above or to the side of the counter, moving the count forward by one. A separate reset pin for each of the two figures on the counters makes it easy to correct a counting error immediately. This

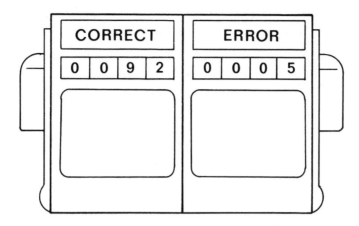

counter is not generally considered to have the figure capability or easy reset qualities that would make it useful for academic tasks. It is limited to a two-figure capability, making 99 the upper count limit before reset; and, because the count button is so small, is not manipulated as quickly or easily as some of the counters that follow.

It is, however, an excellent device for use throughout the day, counting social behaviors. It allows the teacher to keep an ongoing cumulative record of behavior occurrence that can be referred to at any point. There's little danger that the count will get lost, as it easily does on a scrap of paper. The count goes wherever you do, so you don't have to run to a distant corner of the room to mark a behavior on a tally sheet. And, at the end of the day, the count total is already there, without the time-consuming addition of tally marks. Just as important, the wrist counter allows you to count quickly, accurately, and *discreetly*, without making it obvious to the pupil or his peers that you are keeping a count.

Hand tally and dual-channel manual counters. These counters are larger, heavier devices which can be carried in your hand from desk to desk or left at the counting site. They are sturdy counters with big, easily activated pushbuttons and number banks that allow recording of a great number of behavior occurrences before resetting. Their high count capability and ease of manipulation make them perfect tools for quick and accurate recording of lengthy academic assignments, or any high-rate behavior. The counting units are easily combined, allowing you to track a number of behaviors simultaneously, and even to record a variety of components of one behavior—such as correct and incorrect responses—if you choose.

Grocery store counters. Made of brightly colored plastic and designed to register four digits, grocery store counters are often chosen to record academic data. They can be easily carried and manipulated by teacher and pupil alike. Their low cost, attractive colors, and easy use make them especially popular for teachers of grade-school-aged youngsters who want to introduce their pupils to self-counting.

Teacher-made counting devices. Don't overlook the capabilities of teacher-made counting devices. Your opportunities to engage in "creative counting" are limited only by your imagination. Material costs are always a crucial factor in choosing counting implements; and, while the devices just described are well suited to classroom counting needs, it is possible to obtain equally accurate data at far less expense by making your own counting tools. The devices you design yourself are not only easier on your budget, but often much better suited to the unique demands of your own measurement situation.

Creating your own counting devices is easy if you use inexpensive, easily-found materials from kitchen shops, hardware stores, or variety stores. You can design and construct literally hundreds of funny, fanciful, and at the same time, functional counting tools:

— Preschool and elementary teachers find that a kitchen or shop apron, or even a carpenter's tool belt, and small items such as poker chips, nuts or bolts, clothespins, buttons, or dried beans make excellent social behavior counters. The apron should have two (or more, if you need them) pockets. Fill one with the chips, etc.: This is your counting reservoir. Each time the behavior you have designated occurs, transfer a chip from the reservoir pocket to a counting pocket. The tool is portable, easy to use, and functional as well for carrying scissors, pencils, pens, and all the other materials you may find yourself needing throughout the day. (For some children, caution regarding the size and nature of the counters will have to be exercised.)

— A standard pegboard becomes a colorful and functional counting tool, by placing a peg in a hole with each occurrence of the specified behavior. Lacking a pegboard, you may use other materials (e.g., toothpicks and a ball of modeling clay if it's not a safety problem for your children).

— Other materials useful in creating counting tools include: straws, pipe cleaners, bottle caps, cotton swabs, stickers, styrofoam packing chips, and marbles, buttons, or beads.

Knitting counters, purchased inexpensively at sewing and needlework stores, are easily adapted to the demands of classroom counting. They can be fixed atop a pencil for use by the teacher or the pupil himself, or turned into tie clasps, bracelets, and necklace pieces as a functional part of everyday wearing apparel.

The abacus, available in a variety of sizes and materials, is another device easily adapted to counting an endless variety of social behaviors. Simply flip a bead to one side each time a pinpointed behavior occurs. You can keep the abacus at your own or at the pupil's desk, or make it portable by fixing it onto a necktie, belt, or watchband.

Tally sheets remain a counting favorite for teachers of pupils at all ages. Simple to create and use, inexpensive, and open to a wide range of individual designs and applications, the tally sheet is merely a piece of paper, on which you may wish to note information such as the pupil's name, the behavior (or behaviors) you are counting, the date, and the time period during which observation takes place. Leave the tally sheet at your own desk, or the pupil's, or attach it to the wall at some convenient place in the room. (For counting identical behaviors which occur on the part of several pupils during seat work or group discussion time, a tally sheet in the form of a seating chart makes an excellent counting device, with marks placed in each youngster's seating square.) Every time you observe an occurrence of the behavior you are counting, make a tally mark on the sheet. For later ease of counting, teachers often group counts by five.

"Countoons" are a special kind of tally sheet that have been extremely successful in getting pupils to participate in counting their own behavior. They have been applied to a broad range of learning populations (handicapped to gifted), and to a broad range of behavioral issues as well.

Essentially, a countoon is a cartoon that counts. Although countoons have been most frequently applied in relation to social behaviors, they can help youngsters count and change their own behavior across numerous developmental and curriculum content areas. Countoons work for both positive and negative behavior targets: teachers have used them successfully to eliminate a variety of routine classroom behavior problems (e.g., *out-of-seats, hitting, tattling, talking out*), as well as to increase the occurrence of a similar range of desired behaviors (e.g., *speaking clearly, initiating posi-*

tive social interactions, making eye contact, raising hand for attention). Countoons have even been employed to encourage youngsters to work with difficult "inner" behaviors, such as *increasing positive feelings* about themselves, or *coming up with new ideas*. Figure 6:2 contains several examples of commonly used countoons.

FIGURE 6:2. *Countoons Examples.*

Countoons generally have three basic components: (1) the "Behavior" column; (2) the "My Count" column; and, (3) the third, optional, "What Happens" column. A picture illustrating the behavior is found in the first column of the countoon. This picture helps the pupil identify what it is in the first column of the countoon. This picture helps the pupil identify graphically what it is that's being counted. It may be made up of stick figures or other types of cartoon figures, or even a photograph of the youngster engaging in the pinpointed behavior. Children often enjoy supplying this part of the countoon themselves, and their renderings of the behavior in question frequently give teachers provocative insights into how the youngster views his own actions in the classroom. Sometimes, depending upon how the teacher goes about pinpointing, the behavior may be shown as a *movement cycle*, with all three parts of the cycle illustrated (Figure 6:3).

FIGURE 6:3. *Countoon Showing Movement Cycle.*

The countoon's second segment—"My Count"— allows a space for the youngster to make note of his behavior each time it is observed. Although the teacher may have to call the behavior to the child's attention—especially in the beginning—children are often eager and reliable self-counters. Teachers who are concerned about the accuracy of a count can always maintain a separate reliability tally on the side. Children keep their count by making tally marks, just as the teacher would, using a conventional tally sheet. Or, they can simply circle prewritten numbers, as in the count segment shown in Figure 6:4. At the end of the counting period, teacher and child can go over the count together.

The only required components of a countoon are the "What I Do" and "My Count" segments. Some countoons feature a third segment, illustrating the consequence that occurs each time the pupil reaches a specified behavior count. This serves as a handy reminder of what has been decided upon by teacher (or teacher and pupil together) as a fair consequence for a pinpointed problem. Although a teacher may find this column necessary, often encouraging the child to participate in counting is, by itself, enough to change behavior.

These relatively simple devices go beyond their primary counting functions to fulfill some rather complex ends. Countoons create self-awareness. Even without attaching the external contingencies described in a "What Happens" column, they encourage children to make important discriminations about their own appropriate and inappropriate behavior. When youngsters are asked to share in the counting process, teachers are frequently surprised to find that children didn't actually

FIGURE 6:4. *Count Segment of a Countoon.*

My Count				
1	2	3	4	5
6	7	8	9	10
11	12	13	14	15
16	17	18	19	20
21	22	23	24	25

realize what the problem behavior was—even though they had been repeatedly reminded or scolded about it—until they actually saw it illustrated on the countoon. As they were encouraged to participate by watching out for the behavior and counting it themselves, these children gradually became sensitive to when a problem was occurring or was about to occur, and were better able to control their own behavior.

Not only are countoons successful in changing children's behavior, they even work with adults. Figure 6:5 contains a countoon designed by an enterprising ten-year old who wanted to count his mother's behavior. He complained that too often she treated him "like a baby." After thinking about it, he decided that an important first step in changing the situation involved getting his mother to call him "David," instead of "Davey," as she normally did. So he drew up a countoon (Figure 6:5) which showed how he felt when she addressed him as "Davey," and counted his mother's *talk right* behavior. The countoon, taped to the refrigerator in David's home, was a resounding success.

Timing Devices

In addition to counting responses, the timing of some aspects of pupil performance may be an integral part of data collection. Of the devices described in the following pages, some are already a part of your daily activities in or outside the classroom. Others may require purchasing new materials or adapting materials at hand. Whatever timing implements you choose, keep in mind price, availability, and suitability to your own particular needs.

You may want to purchase one or more of the following standard timing devices for a variety of uses.

Classroom wall clocks, digital clocks, and even alarm clocks are, in addition to being standard equipment in most classrooms or homes, easily adapted to a number of timing chores. The cheapest and smallest alarm clock provides a portable, simple to use, and hard to ignore reminder of the conclusion of a timing period. The classroom wall clock is large enough to be seen from any part of the classroom and is, therefore, accessible for timings, no matter where in the room you are working. The major advantage of the digital clock is that it can be used to manage and record timings

FIGURE 6.5. *David's Countoon.*

even by children who have not yet learned to tell time on the conventional wall clock. Any youngster who can copy digits can use a digital clock to record his timings or those of a peer.

Unfortunately, all of these standard timing implements suffer a major drawback. None of them are really suited to short timings: the digital clock, because it lacks a sweep hand; and the wall and alarm clocks, because there is difficulty in monitoring the minute or sweep second hand from a distance while endeavoring to carry on other classroom management activities. Imagine yourself trying to follow an oral reading sample, watching for substitutions, repetitions, and omissions, while keeping your eye pinned to the minute hand of the wall clock at the front of the room! It's a dizzying prospect.

Mechanical timers are available in a wide range of shapes, sizes, and capabilities through most department and variety stores. Plastic kitchen timers, with large setting dials, are colorful, inexpensive, easy to read and easy to manipulate even for youngsters with mild motor coordination problems. They are most accurate for timings of five to sixty minutes. A fleet of these handy timers can do wonders in allowing children to carry out their own timings—encouraging the development of self-management procedures and, at the same time, freeing you for other tasks.

For children with problems discriminating the kitchen timer's short single ring during task involvement, a longer-ring timer may be desired. Or, if it's a longer timing period you want—perhaps for monitoring social behaviors, rather than academic work samples—you may choose to use one of the small parking timers which have setting capabilities of five minutes to two hours. These timers are of perfect size for carrying in your pocket or wearing on a chain, making it possible for you or a pupil to transport the timer to a variety of settings throughout the day. Again, a main disadvantage of these timers is inaccuracy or inconvenience for use in very short (e.g., one-minute) timings.

The stopwatch is your best bet for short timings. It may be more expensive, and more fragile than the options just considered; but it offers several features that compensate for these drawbacks. It has the precision necessary for duration, latency, and rate data, as well as reset or cumulative timing capabilities. The stopwatch is also easily portable. You can wear it on a cord around your neck, making it possible to "transport timings" anywhere in or outside the classroom. Because it is easy to read and to carry, you are free to time at a particular pupil's desk while still watching other classroom activities, or to time several samples simultaneously, moving from desk to desk to give feedback and encouragement.

You have just surveyed some of the standard devices used for classroom timing; but, of course, lots of other options are available. Be as creative as your needs and circumstances allow:

Cassette recorders and tapes. Using cassette recorders and tapes can give you a great deal of latitude in delivering prerecorded instructions and timings. Simply record task instructions on the tape at your convenience, along with a preprogrammed timed interval (made using a stopwatch or other device). A child on a self-managed math program might pick up his own math facts worksheet and bring it, with a tape recorder, to his desk or other appropriate work area. By teaching him to use the *start, stop,* and *rewind* buttons, you enable him to conduct a timing completely independently. Pushing the play button, he hears a prerecorded message giving him the directions he needs and instructing him to "Please begin . . . ." When the desired timing period has elapsed—say, one minute—the tape signals him with a beeper or verbal direction to stop. Once the child can manage to rewind the tape, he is ready to take as many individualized timings as he needs in any subject, at any time during the day.

Clock stamps and cardboard clock dials, purchased through school supply catalogues, make it possible to indicate desired timings on an individual basis for your students. Clock stamps are handily used in connection with worksheets or books, drill materials, or probe materials on which a timing is to be carried out. A blank clock face is stamped on the material, to be filled in by either student or teacher with clock hands showing the time at which work is to begin. An additional empty clock face may be stamped on the work, to be filled in appropriately at the completion of the timing. Similarly, two cardboard clocks with adjustable hands can be used; the youngster matches the time registered on the wall clock as he begins work on one clock, and then registers the time he stops work on the other, making it possible for you to record and use this information later at your convenience.

Music box toys, "hour-glass" egg- or game-timers, and electronic devices used to automatically turn on lights or other household appliances are only a few of the other implements you can modify to fit your own classroom timing requirements. Scrutinizing toy cupboards and catalogs, as well as your own kitchen, garage, or workshop, will bring to your attention other innovative ways to carry out your classroom timing activities with a maximum of ease and enjoyment.

Data-Collection and Storage Forms

Whenever you finish counting and/or timing for the day, you are left with what are called "raw data." These performance scores, behavior counts, or timings—accumulated during whatever sampling procedures you have conducted—are unprocessed, untranslated, and undigested bits of behavior information. If you do not find some way to summarize, compile, and store them for later comparison and interpretation, they'll go on piling up day after day in stacks and stacks of worksheets, exercises, and tally sheets. Charting, one way to store data, is discussed in the next chapter. But there are other ways to keep your raw data under control.

Raw data sheets. A raw data sheet is simply that: a sheet of paper on which raw data are entered and kept in one place. Such a data sheet may be designed to record performance over several weeks or months. It can hold data regarding social behavior counts, or academic performance. Figure 6:6 contains a simple raw data recording form.

FIGURE 6:6. *Raw Data Sheet.*

Name: _____

Behavior: _____

Time: _____

10/5 – 10/9					
10/12 – 10/16					
10/19 – 10/23					
10/26 – 10/30					

Raw data sheets can be designed to hold as little or as much information as you desire. Figure 6:7 contains a much expanded form. Notice the identification information included at the top of the form. Spaces are provided for the names of the child, teacher, and school, as well as for descriptive information about target behaviors, and the beginning and ending of the program. In order to make scoring information as detailed as possible in a small amount of space, the teacher has included columns for entering each performance date, the amount of time spent on task, the number of opportunities for performance (in this case, the number of words on reading practice lists), the correct amount, and the error count. In an additional column, marked Comments, the teacher enters information about program changes, such as the introduction of a new list, or a contingency for correct responses.

Why take the extra step of recording this information—which may already be on a follow-along sheet the teacher uses during the sample each day—on a permanent record form such as this? Just consider, if this teacher takes 30 such samples a day, collecting information on both correct

FIGURE 6:7. *Expanded Data Collection Form.*

Name. . . . .   Behavior . . . .

Teacher. . . . .   Date Begin. . . . .

School. . . . .   Date End. . . . .

Date	Time on Task	Total No. Opportunities	Number Correct	Number Error	Comments

and error performance, she's accumulated a total of 60 pieces of information each day. A daily summary sheet gives the teacher all the behavior information she needs to know about a child's growth for an entire quarter on one sheet of paper. If she wants to see how the child was doing three weeks ago, she doesn't have to thumb through two dozen pages. The teacher has easy and immediate access to the performance of any pupil in her classroom by simply glancing through a raw data summary form.

CONCLUSION

Where data collection is concerned, the Boy Scout motto applies a thousand-fold: Be Prepared. Teachers who rush into the classroom with stopwatch and counter in hand, but no plan in mind, find to their dismay that small details make the difference between reliable data, easily obtained, and a collection of useless numbers. Such details often seem unimportant until we find ourselves embroiled in them. The amount of time it takes to prepare pre-counted work sheets, or to set up a prerecorded taped timing, pays off dramatically in freedom from drudgery during both class time itself and during precious planning periods. The planning and organization such activities require are minimal in comparison to their benefits. They free you during class time for a flexibility and involvement in classroom activities that enhance the quality of all your efforts.

Chapter 7
Measurement Plan Sheets

Instructional plans are blueprints for action in the classroom. They give teachers an opportunity to outline strategies in advance; to rehearse and analyse moves before they are made; and to visualize the size, shape, and direction of things to come. But, while most teachers admit the importance of making plans, few teachers enjoy writing them out. Plan writing can be a tedious business, regarded by many as having only a tenuous relationship to the dynamics of classroom interaction. Some teachers claim that formalizing instructional plans interferes with their own personal teaching style. Entirely at ease working from mental blueprints, they feel that writing a plan down puts unnecessary restrictions on the freedom and spontaneity with which they conduct lessons. Others dread plan writing, but do it anyway because they are required to by school administrators.

Realizing the depth of bad feelings about plan sheets, why introduce a chapter devoted to them? It is because we are certain that the planning strategy proposed here serves as something larger than a mere device for recording lesson plans. Whether or not you decide to implement one of the three plan sheets introduced in this chapter, you can still put any one of them to excellent use as a framework against which to evaluate your present planning efforts. The issues these plan sheets raise are central to any good instructional plan, and can help you analyse some of the difficult planning problems you are likely to confront. Simply reading the questions a measurement plan asks teachers to answer about the instructional environment, will encourage you to consider new and important aspects of the learning world you are creating for each child in your classroom.

If you can, suspend your antipathy temporarily, and explore with us what we call "the plans you'll love to hate."

PLAN SHEETS FOR PROBLEM SOLVING

The plan sheets that follow are basically three variations upon a single strategy for analyzing and describing the details that go into an instructional plan. Beginning with an extremely simple, three-column format, these plan sheets grow progressively more complex in order to make possible a more precise description of the learning environment. Before getting into the plan formats themselves, let's consider briefly some of the general advantages to having a written plan.

First, for all the talk of plan writing being time consuming, it can actually save hours of valuable instructional time and avoid later loss of momentum on the part of both teacher and pupil. It gives us something concrete to *analyze before* we actually find ourselves in the classroom; many times this allows us to spot structural weaknesses before a plan is implemented. Further, a written plan gives us a reference point, something to *analyze during* the course of instruction to ensure that nothing important is being left out, and nothing extraneous being added. Finally, a written plan serves as a *permanent record* of what has taken place in a very complex, rapidly changing environment. Writing a plan down leaves a kind of history of the child's learning world, from which better plans for future instruction can be drawn.

Specific Advantages of a Measurement Plan

In addition to these general advantages, the plan sheets offered in this chapter carry some extra pluses.

Major focus on the child's behavior. Traditionally, plan sheets have geared teachers to concentrate mainly upon their own part in the instructional equation. For instance, they might spell out in painful detail the particulars of lesson presentation—the introducing, explaining, directing, and cueing activities commonly thought of as "instruction." Relatively little emphasis has been directed toward specifying what the *child does* to demonstrate learning, or a failure to learn. Since the emphasis of this book is upon behavior as a measure of learning, it is natural to look for a plan sheet which not only incorporates behavior, but focuses upon it.

In all of the plan sheet variations presented, behavior takes center stage, both visually and symbolically. The central column on each sheet is labeled Behavior, and is designed to carry a description of the behavior being measured, as well as details about the way in which measurement will be accomplished. This column's location on the plan sheet highlights behavior's primary position in any teaching plan: The child's behavior is seen as the pivot of a seesaw, with the teacher's presentation of instructional material and subsequent response to child performance as left and right sides which move and balance in relation to the behavioral fulcrum.

Identification of critical environmental influences on child performance. A dominant function of this type of plan sheet is to help teachers identify critical factors in the environment which affect child performance in the measurement situation. The variables a teacher hopes to identify and isolate are those which exert consistent, ongoing influence, regardless of whatever transient sorts of happenings might be taking place (e.g., It's Monday; the child's just had an argument with Mom; some of his peers wouldn't let him join them at recess; he's going to visit a favorite aunt after school).

To aid teachers in isolating these factors, the plan sheets presented each offer a basic format which serves as a guide in isolating potentially critical elements in the environment; not only instructional procedures (the presentation part of the plan), but also whatever plans exist for proceeding after the child has performed.

Incorporation of measurement details. These plan sheets have been designed especially to be used in conjunction with classroom data taking. Measuring performance requires a few steps which may be new to teachers: choosing a performance measure; specifying data collection (counting and timing) tools; making arrangements for data collection in the course of instruction (e.g., deciding when, where, and for how long). The new plan format provides a place where each of these details can be organized and spelled out ahead of time. This encourages teachers to think through data-collection details before they are incorporated as a part of actual classroom procedures, ironing out some of the wrinkles before they have a chance to embed themselves in the fabric of the plan.

A plan sheet designed to last. Plan writing is aversive to teachers for a number of reasons. For one thing, they find themselves torn between writing a plan that looks like *War and Peace* (every detail is there, from the first breath to the last gasp), and one that resembles a Western Union telegram (so brief it may be impossible to decipher later on). Once the plan's written, there's often a need for constant updates and rewrites.

The planning strategy presented here attempts to address both these problems. First, it helps teachers strike a balance between the complete and the concise, weeding out inconsequential detail and yet leaving a plan that contains all the essential information. What we call planning guidelines, or master plans, accompany each of the three plan sheet variations. These planning guides contain a list of questions under each of the major column headings. The questions are designed to help the teacher think through the plan quickly and logically, identifying what seem likely to be the most critical, essential elements for that particular plan. Writing the plan is simply a matter of entering the answers to these questions, as briefly as possible, within the appropriate spaces.

The writing task is further lightened by the use of what we call "plan sheet shorthand." Teachers are encouraged to develop and use a consistent set of abbreviations, codes, and catch phrases (e.g., "T" for "The teacher"; "Lipp. 5" for *Lippincott Basic Reading Series*, Level 5; "S" for "Student," or a letter abbreviation for the child's name). The aim is a plan that provides—rather than a blow-by-blow description complete with every "and," "this," and "the"—a skeleton which the teacher fleshes out as he implements the plan in the classroom.

Most important, the planning strategy encourages teachers to write plans which *identify general instructional patterns in the learning environment that remain constant over several days, weeks, or even months.* This is what we mean by a plan that is designed to last. The teacher develops a basic plan framework which does not change constantly from one day to the next. It is intended to stay in place long enough for the teacher to determine which elements of the plan are beneficial and successful and which need modification. The plans are designed to be used in conjunction with data collected on child performance: a powerful combination. When data suggest that an initial plan is failing, the teacher need not abandon the total plan. Instead, gradual changes are made by analyzing all the plan components and changing first one, then another, to determine which parts of the plan can be salvaged and which should be discarded in favor of new strategies.

Through this process of systematic analysis and change, the teacher is able to obtain critical information about certain dimensions of the plan which may have generality to other plans for the same child (e.g., certain modes of presentation are most effective; or the way in which feedback is delivered proves significant). Teachers who use plan sheets to increase their analysis capabilities continue to be amazed at their growing abilities to create and implement new ideas and strategies tailored to meet individual instructional needs.

Plan Sheet 1: The Before-After Plan Sheet

The Before-After plan sheet (Figure 7:1) provides an extremely simple, three-column format for outlining important instructional variables. Its columns, labeled Before, Behavior, and After, encourage the teacher to enter critical instructional events on the plan sheet in a sequence which progresses from left to right across the page. This sequence replicates, for the most part, the sequence of events that actually occur during instruction. Placing instructional events within a before/after context in relation to behavior also clarifies the distinction between variables which might trigger or induce behavior (the Before), and those which might reinforce or encourage behavior to continue (the After).

FIGURE 7:1. *Before-After Plan Sheet.*

Plan Sheet Number . . . . Location. . . . . .Date . . . .

Pupil Name . . . . .Chronological Age. . . . . Teacher . . . .

Pinpointed by . . . .

BEFORE	BEHAVIOR	AFTER

FIGURE 7:2. *Plan Sheet Heading.*

Plan Sheet Number ..*1*............... Location..*Language Arts*..Date *10-6-86*........

Pupil Name ..*Danny Osborne*..Chronological Age. *16*........... Teacher *Mark Zeldin*

Pinpointed by ..*Zeldin & Katagiri*.......................................

To make the plan sheet's format easier to understand, read how one teacher—Mark Zeldin, a Language Arts and Literature teacher at Jefferson High School—uses it to describe the group discussions he conducts on assigned readings in an eleventh grade American literature unit. Each of the plan sheet's components are discussed separately.

Plan sheet heading. Running across the top of the plan sheet are spaces for information designed to simplify the task of identifying, sorting, storing, and relocating plans. Figure 7:2 shows this portion of the plan sheet as it is filled out for a particular student in Mark Zeldin's classroom. According to the information recorded here, this is the first of an as yet undetermined number of plan sheets that Mr. Zeldin will devise throughout the semester for his eleventh grade Language Arts class. The plan is written and implemented on October 6, 1986.

Although the presentation techniques described on the plan sheet apply to the entire class, this plan has been written up specifically for sixteen-year old Danny Osborne who already looks like he will fulfill the reputation that has traveled with him. Danny is no troublemaker, but he has frequently been described by past teachers as "withdrawn" and a "loner." Mr. Zeldin observed that, indeed, Danny rarely contributes to class discussions, and then only when he's pinned down. Mark Zeldin intends to track Danny's class participation, using the plan sheet presented here. Because he consults with the school psychologist, Faith Katagiri, who is familiar with Danny's school record and test results, Ms. Katagiri's name appears with Zeldin's on the plan sheet in the "Pinpointed by" slot.

The Before column. The plan sheet's Before column contains a description of all those events referred to as "antecedent," or happening before the child engages in the pinpointed performance. Typically, it includes the presentation details of the plan, such as mode of presentation (whether verbal or nonverbal), the type of demonstration used, any special instructions, whether modeling is provided, practice allowed, as well as a description of instructional materials (e.g., textbooks, dittos, workbooks, concrete manipulative materials). Other pieces of information which may be relevant include when the program occurs in the day and how often it occurs during the week.

Figure 7:3 shows how Mark Zeldin has filled out this column to describe his general teaching plan. The American Literature discussion group takes place daily between 10:20 and 11 o'clock. Zeldin's presentation follows a fairly consistent pattern. He begins by offering a brief review of the piece of literature to be covered for the day. In the course of the presentation, he provides biographical information about the author, adding his own observations about the period in which the particular piece was written. He presents setting and characters, and goes on to discuss major themes that arise. He then goes on to explore character conflicts, insights, and growth. Finally, Zeldin examines the author's style and how it functions to communicate explicit and implicit messages within the work. He may compare and contrast this work's treatment of a particular theme with other

FIGURE 7:3. *Before Column.*

BEFORE	BEHAVIOR	AFTER

American Literature Discussion 10:20 to 11:00
 	
Novel or short story to be covered: *Sleepy Hollow* by Washington Irving
 	
Before presentation, tell s. they should take notes and raise hands if they have questions or comments while T. is speaking.
 	
  1. Begin with brief review of work
      a. Biographical information about author
      b. Observations about period in which work was written
 	
  2. Present setting and characters
  3. Discuss major themes and author's treatment
  4. Explore character conflicts, insights, growth
  5. Examine style
  6. Compare and contrast this with other works; other authors' treatment of similar themes		

works covered in the class and/or with previous authors' treatments of similar subject matter. Before the presentation begins, Mr. Zeldin encourages students to take notes and raise their hands if they have questions or comments as he's speaking. The story listed for the week's reading and discussion is Washington Irving's *Sleepy Hollow*.

    The Behavior column. In the Behavior column (Figure 7:4), all pertinent details relating to the definition and measurement of the pinpointed behavior are specified. For instance, along with the pinpoint, the teacher may include a behavioral objective if one has been determined. Mark Zeldin defines two behaviors (*asks related questions* and *makes related comments*) which he counts and weights equally. Zeldin feels that individual participation in the form of questions and comments is one of the easiest and most direct measures of each student's grasp of the material. The objective, dated for the end of the year, is three or four comments and/or questions during each discussion.

Zeldin also outlines information related to measurement. He decides to count Danny's responses using a tally sheet attached to daily discussion notes, recording results in terms of raw score in the gradebook at the end of the day. (Ms. Katagiri has already recommended that he consider refining his measurement system with the introduction of a behavior chart for the data he has been accumulating.)

The After column. Events which happen After, or subsequent to performance, include feedback in the form of praise, comments, correction, and further instruction. Mr. Zeldin's entries in this column (Figure 7:5) indicate that whenever a student has a question or comment, he stops his own discourse to call on the pupil, making relevant comments or giving answers based upon the student's contribution.

FIGURE 7:4. *Behavior Column.*

BEFORE	BEHAVIOR	AFTER
American Literature Discussion 10:20 to 11:00	Makes related comments Asks related questions	
Novel or short story to be covered: *Sleepy Hollow* by Washington Irving	Objective: 3 to 4 comments and/or questions per class period; by the end of the year	

Before presentation, tell s. they should take notes and raise hands if they have questions or comments while T. is speaking.
 	
1. Begin with brief review of work
  a. Biographical information about author
  b. Observations about period in which work was written
2. Present setting and characters
3. Discuss major themes and author's treatment
4. Explore character conflicts, insights, growth
5. Examine style
6. Compare and contrast this with other works; other authors' treatment of similar themes
	T. counts, using tally sheet; records results as raw score in grade book at day's end	

FIGURE 7:5. *After Column.*

BEFORE	BEHAVIOR	AFTER
American Literature Discussion 10:20 to 11:00	Makes related comments Asks related questions	T. comments or answers s. after calling on him.
Novel or short story to be covered: *Sleepy Hollow* by Washington Irving	Objective: 3 to 4 comments and/or questions per class period; by the end of the year	
Before presentation, tell s. they should take notes and raise hands if they have questions or comments while T. is speaking.	T. counts, using tally sheet; records results as raw score in grade book at day's end	

1. Begin with brief review of work
   a. Biographical information about author
   b. Observations about period in which work was written
2. Present setting and characters
3. Discuss major themes and author's treatment
4. Explore character conflicts, insights, growth
5. Examine style
6. Compare and contrast this with other works; other authors' treatment of similar themes		

Mr. Zeldin reviews the data with Ms. Katagiri two weeks after initiation of Plan 1. They agree that a plan change is in order. Danny contributes very little. Ms. Katagiri suggests that the plan itself is probably largely responsible. Especially important is the observation that Zeldin has structured the discussion situation so that very little participation is required from any but the most motivated students. Judging from the Before column, it appears that Zeldin carries the discussion singlehandedly almost every day. Although his presentation is carefully planned and artfully executed, there is little evidence that it is designed to encourage student contributions. The easiest and most obvious place to start seems to be with a restructuring of the Before part of the plan. One of the first changes Ms. Katagiri suggests is that Zeldin make a concerted effort to solicit questions and opinions throughout the presentation each day. If this alteration in the plan doesn't have the desired effect, the teacher can investigate changes in other components of the plan sheet.

   A Before-After planning guide. To make filling out your own Before-After plan sheets easier, a planning guide, or Master Plan (Figure 7:6), is provided which includes some of the important questions you'll want to ask yourself about plan implementation.

FIGURE 7:6. *Before-After Planning Guide.*

Plan Sheet Number . . . . Location . . . . Date . . . .

Pupil Name . . . . Chronological Age . . . . Teacher . . . .

Pinpointed by . . . .

BEFORE	BEHAVIOR	AFTER
Who's involved Who was present Who presented informa- mation, instructions, directions, material Who chose subject, ma- terial, activity *What's* involved What material, directions, etc. and any specific de- tails about it—i.e., title, pub- lisher, level *When* presented Time of day (when in week), Order of activity	Specify behavior Performance measure chosen (e.g., percent, raw score, rate) Data collection tools used Other details regarding data collection proce- dures (e.g., who takes data; how long beha- vior observed during day) Where data recorded (e.g., grade book, chart) Objective (e.g., information about level of performance desired, completion date anticipated)	*Who's* involved Who is present; who delivers feedback *What's* involved Note home, praise, smiles, stickers, grades, trinkets, food *When* presented When is child to receive a con- sequence for the behavior specified on this plan. How frequently is he to re- ceive this?
Where presented Any potentially significant details about surround- ings or place of presen- tation *How* presented Mode of presentation used Details of presentation considered potentially significant		*Where* presented Where is the feedback, etc. to take place *How* presented Auditory/visual/tactile- kinesthetic; cash-in, trades involved

Plan Sheet 2: The Is-Does Plan Sheet

The Is-Does plan sheet, while slightly more complex than the Before-After plan, has as its goal identification of exactly the same sorts of instructional components. Its five columns are filled out and read in a similar left to right sequence. The material entered in the Program and Programmed Events columns typically constitutes the "Before" part of the plan; Arrangement and Arranged Events entries on the right hand side of the plan sheet describe those instructional components which follow upon occurrence of (After) the specified behavior.

O. R. Lindsley introduced the Is-Does several years ago, and today, after a number of subsequent modifications, variations on the original form continue to enjoy great popularity, especially in classrooms where Precision or Exceptional Teaching procedures are practiced. The name of the plan sheet was derived from the fact that it is designed to help teachers distinguish between instructional events that are merely a part of what *is* happening in the classroom, and those events that are a part of a plan which eventually *does* work as the teacher intends it to. The first plans a teacher writes are usually of the *is* nature: they describe what is going on in the child's program with no claims about what works. Eventually, after systematic modifications of individual plan components the teacher arrives at a *does work* plan.

Figure 7:7 contains a blank Is-Does plan sheet. Let's fill it in, piece by piece, using components of a program designed by a Junior High social studies teacher to change the "talking out" behavior of one pupil in her Current Events class.

Plan sheet heading. The identification information required on the Is-Does plan is identical to that found on a Before-After plan sheet form. On this particular plan sheet the teacher, Nancy Jones, has entered the name of the offending party, Andy Knapp, as well as his age and the number and date of the plan.

The Program and Programmed Events columns. Corresponding roughly in content to the Before section of the Before-After plan sheet are the first two columns on the left side of the Is-Does plan, labeled Program and Programmed Event. In the first of these, the Program column, are entered such program details as time, name of activity scheduled, text title, level, publisher, and/or a description of other instructional materials which may be employed. The program's starting date is also included. Ms. Jones has duly noted this information on her plan sheet (Figure 7:8). The current events discussion she conducts occurs daily from 9:00 to 9:20. All 30 children in her social studies class participate in the discussion, which ranges over a variety of topics selected by Ms. Jones from current newspapers and news magazines, as well as recent network radio and television broadcasts.

The second section of the Is-Does plan sheet is labeled Programmed Event. In this column are outlined all those environmental components planned, or programmed, to be presented regardless of whether or not the child engages in the pinpointed behavior. These events include directions, demonstrations, and other details describing the way in which instructional material is presented by the teacher, by other students, or by audio-visual devices. In most cases, programmed events take place prior to pupil performance.

At the beginning of each session, Nancy Jones informs the group that they should raise their hands to ask questions or make comments. She reminds them that, even though she knows how eager they are to get in on the discussion, she wants them to wait until the individual speaking is finished before presenting their own views. The Programmed Event column in Figure 7:8 shows how Ms. Jones records these instructions on her plan sheet.

Plan Sheet Number . . . . Location. . . . . .Date . . . .

Pupil Name . . . . .Chronological Age. . . . . Teacher . . . .

Pinpointed by . . . .

PROGRAM	PROGRAMMED EVENT	MOVEMENT CYCLE	ARRANGEMENT	ARRANGED EVENT
Title plus times dates	Presentation criteria, i.e., who, what, how	Record how long, by whom, how	By whom ratio or interval	Presentation criteria, i.e., who, what how

The Behavior column. The Behavior column primarily specifies and describes the pinpointed behavior. If both correct and incorrect response components are considered, each is carefully defined. In addition, details concerning data collection are worked out and entered in this column on the plan sheet before the plan is initiated, making data gathering much easier and more efficient. The amount of time over which the pinpointed behavior will be observed and counted is one of these data taking issues. Recording time is especially important if it is not the same as program time. (For instance, in the case of reading instruction, a reading group may last from 10:00 to 10:30, but individual performance data on the reading of controlled vocabulary lists is collected using one-minute samples.)

Ms. Jones defines Andy's frequent outbursts as *interrupts* (Figure 7:10). As they are likely to occur at any time during the 20-minute discussion, she counts throughout the entire session. The teacher uses a golfer's wrist counter because she finds it easier and less obtrusive than other methods. She knows that if she were to use a tally sheet, for example, she would be obliged to transport it with her as she moved about the room, or continually return to her desk to record a tally mark each time Andy interrupted. With a wrist counter, Ms. Jones can quickly and discretely punch in a count

Plan Sheet Number *1* . . . . Location *Current Events* Date *10-4-84* . . . .

Pupil Name *Andy Knapp* Chronological Age *14* . . . . Teacher *Nancy Jones*

Pinpointed by *Jones* . . . .

PROGRAM	PROGRAMMED EVENT	BEHAVIOR	ARRANGEMENT	ARRANGED EVENT
10/4/84 9:00−9:20 Current Events Discussion Class of 30 M−F T. choice topics				

after each interruption no matter where she happens to be standing, or what other teaching activity she is engaged in. She charts the behavior count on an arithmetic graph. Her eventual goal is for Andy to make it through the entire 20-minute period without interrupting at all.

The Arrangement and Arranged Event columns. The final two columns, located on the right side of the Is-Does plan sheet are labeled Arrangement and Arranged Event. We'll deal with these two columns together because they are used in combination to describe the way in which planned events follow upon behavior.

The information found in the Arranged Event column is similar in nature to that recorded in the After section of the Before-After plan sheet, but with certain very important differences. While *all* events subsequent to behavior may be outlined in the After portion of the Before-After plan, the events described under Arranged Events must be contingent upon performance; that is, they have been *specifically planned and arranged* by the teacher to occur only if the specified performance occurs. This distinction is easily understood with the help of a few examples. A teacher using the Before-After plan sheet may note a wide variety of events which follow performance, but which are not contingent upon it. For instance, that class is dismissed at the end of the math period, whether or not the child has finished the assignment. This is an After event which occurs regardless of the

FIGURE 7:9. *Programmed Events Column.*

Plan Sheet Number *1* . . . . Location *Current Events* Date *10-4-84* . . . .

Pupil Name *Andy Knapp* . Chronological Age. *14* . . . . Teacher *Nancy Jones* .

Pinpointed by *Jones* . . . .

PROGRAM	PROGRAMMED EVENT	BEHAVIOR	ARRANGEMENT	ARRANGED EVENT
10/4/84 9:00–9:20 Current Events Discussion Class of 30 M–F T. choice topics	T. says, "Raise your hand if you want to ask or answer a question. If you have a comment about something being said, be sure to wait until the other person has finished speaking."			

amount or quality of performance. In other cases, a child may be allowed to participate in a variety of activities which follow performance temporally but are not contingent upon any specified amount or quality of behavior (e.g., when the teacher says spelling period is over, the child is allowed to collect all the papers; he goes to recess; he is blackboard monitor; or he joins the other children in playing a game). An Arranged Event, on the other hand, is one which occurs only when the child performs in a specified way (e.g., he must complete all his math problems to go to recess; he must achieve a score of 100 percent to be blackboard monitor; he must get two errors or less to join in a free time activity).

Before dealing any further with the contents of the Arranged Events column, let's return to the Arrangement column which precedes it. This column is located between Behavior and Arranged Event because it is used to describe the special relationship (or arrangement) between behavior and the events which follow. This is an "if-then" type of relationship; that is, *if* the behavior happens, what *then*? The relationship is stated in terms of a numerical ratio, x:y. The left side of the ratio (x) describes the *if*, and the right side of the ratio (y) describes the *then*. For easy interpretation, just remember that the number on the left represents the amount of behavior that must occur before a consequence is delivered, and the number on the right refers to the number of events arranged to follow upon performance.

FIGURE 7:10. *Is-Does Behavior Column.*

Plan Sheet Number *1* . . . . Location *Current Events* Date *10-4-84* . . . .

Pupil Name *Andy Knapp* Chronological Age. *14* . . . . Teacher *Nancy Jones*

Pinpointed by . *Jones* . . . .

PROGRAM	PROGRAMMED EVENT	BEHAVIOR	ARRANGEMENT	ARRANGED EVENT
10/4/84 9:00–9:20 Current Events Discussion Class of 30 M–F T. choice topics	T. says, "Raise your hand if you want to ask or answer a ques-tion. If you have a comment about something being said, be sure to wait until the other person has finished speak-ing."	Interrupts *Counted* Over 20 minutes M–F By teacher Using wrist counter *Recorded* Arithmetic graph *Objective* 0 per 20 minutes		

The two sides of the ratio may be equal (e.g., a ratio of 1:1, read as "one to one"). Or, they may be unequal, with the left side either larger or smaller than the number on the right. For instance, a teacher may discover that a particular child works much more accurately if each answer on his math assignment is marked with a star in red pencil. The arrangement is described as 1:1. For each correct answer, one red penciled star. Or perhaps the teacher corrects work at the child's desk, with the youngster watching, and wants two things to take place for every correct answer: the teacher marks a star and also gives some verbal feedback, like "Good for you, another one correct!" In this second case, the arrangement would be entered as 1:2; that is, one behavior gets two arranged events. An arrangement of 2:1, on the other hand describes a still different type of situation. Remember that the number on the left side of the ratio refers to the number of behaviors required per arranged event. A ratio of 2:1 indicates that two occurrences of the pinpointed behavior must be observed before a consequence takes place (e.g., for every two problems the child gets correct, the teacher may give one "happy face" sticker).

Innumerable studies by behavioral scientists, and the practical experiences of hundreds of classroom teachers, demonstrate that different arrangements affect and maintain behavior in very marked, consistent ways. These findings may encourage you to seek more detailed information about basic arrangement types (also called schedules of reinforcement). Any good text on behavioral principles will contain such information.

Once the arrangement and arranged event are decided upon, putting them on the plan sheet is quickly accomplished by entering a combination of a numerical ratio in the Arrangement column and a brief description of the events signified by that ratio in the Arranged Event column to the right. Figure 7:11 contains the Arrangement and Arranged Events segments of Nancy Jones' plan

---

FIGURE 7:11. *Arrangement and Arranged Events Columns.*

Plan Sheet Number *1* . . . . Location *Current Events* . Date *10-4-84* . . . .

Pupil Name *Andy Knapp* . Chronological Age *14* . . . . Teacher *Nancy Jones* .

Pinpointed by *Jones* . . . .

PROGRAM	PROGRAMMED EVENT	BEHAVIOR	ARRANGEMENT	ARRANGED EVENT

---

sheet which describe briefly what happens every time Andy interrupts. Nancy says, "Andy, please wait until . . . . (person speaking) has finished talking and then raise your hand." The arrangement is noted as a 1:1; that is, each time Andy interrupts (one interruption) the teacher responds by requesting that he raise his hand if he has something to contribute (one teacher response).

The total plan sheet, shown in Figure 7:12, offers a concise description of the details Ms. Jones considers relevant to the problem she is looking at. If the data she collects show that her plan

FIGURE 7:12. *Completed Is-Does Plan Sheet.*

Plan Sheet Number ..*1*.............. Location *Current Events* .Date *10-4-84* ....

Pupil Name *Andy Knapp* .Chronological Age. *14* .... Teacher *Nancy Jones* .

Pinpointed by *Jones* ....

PROGRAM	PROGRAMMED EVENT	BEHAVIOR	ARRANGEMENT	ARRANGED EVENT
10/4/84 9:00–9:20 Current Events Discussion Class of 30 M–F T. choice topics	T. says, "Raise your hand if you want to ask or answer a question. If you have a comment about something being said, be sure to wait until the other person has finished speaking."	Interrupts *Counted* Over 20 minutes M–F By teacher Using wrist counter *Recorded* Arithmetic graph *Objective* 0 per 20 minutes	1:1	T. says, "Andy, please wait until Joe has finished speaking, and then raise your hand."

is not succeeding in diminishing the overall occurrence of interruptions each day, Ms. Jones can make program changes in any one or more of the five program areas examined on the plan. In Figure 7:13, a plan change is added two weeks after behavior data suggest that Andy's interruptions are growing more, rather than less, numerous each day. Andy's teacher might have decided to alter her instructions or opening remarks in hopes that a different programmed event would have a favorable impact. She decides instead to change the consequences of the behavior. Each time Andy interrupts in Current Events now, she instructs him to make a tally mark on a countoon, recording his own behavior. Although the countoon is a counting device and counting devices and strategies are normally noted under Behavior, Ms. Jones notes the countoon under Arranged Event. She has called counting to Andy's attention as a part of her plan to change his behavior, and it thus qualifies as a consequating event.

The program change is noted simply by drawing a line beneath the first plan, noting the date of change in the Program column, and entering plan alterations in appropriate spaces.

Is-Does planning guide. In Figure 7:14, an Is-Does master plan sheet outlines the major questions which may be asked as an Is-Does plan is constructed.

FIGURE 7.13. *Plan Change on an Is-Does Plan Sheet.*

Plan Sheet Number *142* . . . . Location *Current Events* Date *10-4-84* . . . .

Pupil Name *Andy Knapp* Chronological Age *14* . . . . Teacher *Nancy Jones* . .

Pinpointed by *Jones* . . . .

PROGRAM	PROGRAMMED EVENT	BEHAVIOR	ARRANGEMENT	ARRANGED EVENT
10/4/84 9:00–9:20 Current Events Discussion Class of 30 M–F T. choice topics	T. says, "Raise your hand if you want to ask or answer a question. If you have a comment about something being said, be sure to wait until the other person has finished speaking."	Interrupts *Counted* Over 20 minutes M–F By teacher Using wrist counter *Recorded* Arithmetic graph *Objective* 0 per 20 minutes	1:1	T. says, "Andy, please wait until Joe has finished speaking, and then raise your hand."
10/18/84			1:1	T. asks Andy to make tally mark on "Interruptions" countoon

Plan Sheet 3: The Increase-Decrease Plan Sheet

A slightly more refined version of the Is-Does format allows teachers to track plans for both increase and decrease targets simultaneously. This is called the Increase-Decrease plan sheet (Figure 7:15). The plan format is identical to that of the Is-Does in all respects, except that it contains two sets of Behavior/Arrangement/Arranged Event columns, one for each of the behavior targets tracked.

These targets were described as *fair pairs* in Chapter 2 (e.g., correct and incorrect responses in oral reading; *talking out* versus *raising hand* to speak in a social context). Perhaps you've previously given little or no thought to defining two sides of a behavioral issue. Teachers commonly believe that by devoting all their efforts to increasing the occurrence of desired behaviors, other less

FIGURE 7:14. *Is-Does Planning Guide.*

Plan Sheet Number . . . . Location . . . . .Date . . . .

Pupil Name . . . . .Chronological Age . . . . Teacher . . . .

Pinpointed by . . . .

PROGRAM	PROGRAMMED EVENT	BEHAVIOR	ARRANGEMENT	ARRANGED EVENT
Day: mo/year Time: ___to ___ Title: Publisher Author Book and level General description of material Pages Specify content Specify format

Name of class

Number in class

When in week does subject occur? M T W TH F

When in a day does subject- activity se- quentially occur? 1 2 3 4 5 6 7 8 9 10

Who chose sub- ject-activity? | Manner materials obtained

How presented a. Type of pre- sentation: Instruct/ Demo/As- sign/Other (Specify) b. Mode cho- sen: Auditory/ visual/tac- tile-kines- thetic/com- bination (Specify) c. Materials us- ing mode (e.g., black- board, cas- sette tape, mini-compu- ter d. Practice in- cluded e. Prompting/ priming de- tails f. Instruction included for | Specify behavior a. Define cor- rect b. Define error

Who pinpoints

Who corrects Teacher/pupil/ peer/cross-age/ tutor/aide/ volunteer/ other

General data col- lection issues How long a period is be- havior ob- served/ data col- lected? How frequent- ly are data taken? Who decides, collects data?

Data collec- tion tools | If: Then

Lefthand side of ratio refers to Behavior col- umn (1:1)

Is feedback given for: a. Specified number of occurrences of pinpoint- ed behavior (e.g., speci- fied number of corrects, specified number of errors) b. Specified time spent engaged in behavior

Righthand side of ratio refers to number of ar- ranged events child will re- ceive (1:1) | What feedback presented (in- clude details of correction pro- cedure for cor- rects versus er- rors)

What received (praise, smiles, touches, stick- ers, written words, grades, more work, display, note home, free time, trinket, food)

What cash-in, trade, etc. in operation? (provide de- tails)

Who decides the above?

Who delivers it?

How feedback given? |

-164-

(pupil/peer/other/T/IEP committee) Who chose material to be used? (T/IEP committee/pupil/peer/other) Relation to master or framework for total curriculum plan	general program management (e.g., informing where to find paper, pencil, worksheet; when to start work; how long they have to complete work; details about recurring teacher contact; contact criteria; stopping criteria) What presented? How often presented? Who chooses presentation and who presents	(what tools used—stopwatch, counters, tally sheet) Data records (what records kept, e.g., grade book, sheet, chart) Who decides recording form Who records data Behavioral Objective	How many arranged events offered (e.g., in this case only one) Other: Schedules of reinforcement you might cite: Fixed interval/variable interval/fixed ratio/variable ratio Who decides? Arrangement application (entire program, partial) Correct-error distinction (e.g., different arrangement for corrects than errors) Details of a contract	What mode the person delivering feedback uses to present? Verbal and/or physical/auditory/visual/tactile/kinesthetic/olfactory/gustatory/combination Criteria for moving on to next objective in program or graduating

desirable behaviors will automatically fade away. They assume that as desirable behaviors become a stronger part of the child's repertoire, because they are incompatible with the pinpointed target, there will be little or no room for the undesired behavior to occur (e.g., if a child is writing on his assignment, he'll have no time for talking out, throwing his pencil, or getting out of his seat). In contrast, others claim that attention must be focused directly upon whatever undesirable behaviors make themselves obvious. Once these have been eradicated, positive, more desirable behaviors should begin to appear and flourish. Neither of these viewpoints has proved exclusively correct.

There is significant evidence that both correct and error, or positive and negative aspects of behavior may require attention simultaneously. Even when intensive efforts are focused upon a particular positive behavior, competing undesirable responses do not necessarily disappear. They may

FIGURE 7:15. *Increase-Decrease Plan Sheet.*

Plan Sheet Number . . . . Location. . . . . Date . . . .

Pupil Name . . . . Chronological Age. . . . . Teacher . . . .

Pinpointed by . . . .

		INCREASE OR MONITOR			*DECREASE OR MONITOR*		
PROGRAM	PROGRAMMED EVENT	BEHAVIOR	ARRANGE-MENT	ARRANGED EVENT	BEHAVIOR	ARRANGE-MENT	ARRANGED EVENT

follow a variety of patterns, including simply hanging on as stubborn problems, or even increasing as the desired behavior increases. Not all undesirable behaviors are incompatible all of the time with other, more desired behaviors. Many teachers have encountered the power-house who can increase the amount of correct written work he does and still find time to hit, throw, and engage in other troublesome behaviors. Dealing solely with an undesirable behavior, on the other hand, you may devote considerable time to eradicating it, only to find that nothing positive rushes in to fill the void. The child does less of what you didn't want, but he doesn't seem to know how to replace previous "bad" behavior with anything better.

Since you cannot assume that a child will automatically take up a whole new set of behaviors without concerted teaching effort directed at their acquisition, you may want to define both positive and negative behaviors separately from the beginning, and specify what happens in the case of occurrences of each separately. Placing Behavior/Arrangement/Arranged Event columns for correct and error side by side allows you to compare definitions and plans for each at a glance. No matter what your philosophy about the ways in which each aspect of behavior should be treated (e.g., ignoring one while concentrating on the other), the Increase-Decrease plan provides a helpful perspective from which to examine the choices you have made.

The following are examples of the way in which teachers from vastly different educational settings have attacked a variety of problems using the Increase-Decrease approach.

An Increase-Decrease plan for introducing an initial communication skill. Sheila Duncan is a teacher in a classroom for severely/profoundly handicapped children who uses the Increase-Decrease plan sheet to describe a program designed to increase the receptive communication skills of her pupil, Roy. Her objective is to get Roy to turn his head toward the sound of a human voice.

According to her plan (Figure 7:16), Ms. Duncan sits behind Roy and positions his head at the midline. During a two-minute session, timed with a stopwatch, she gives repeated cues, leaning to one side of Roy's head and then the other, and saying into his ear, "Roy, look at me." The cue is delivered randomly from alternate sides at ten-second intervals. If Roy turns his head to the correct side within five seconds of the cue, he is praised and given a quick sip of orange juice from a squeegee. If he hesitates longer than five seconds, or turns his head in the wrong direction, he is ignored and receives no juice. (Before delivering each cue, the teacher repositions Roy's head at midline.)

Correct and incorrect responses are noted on a mechanical dual tally counter. The aim for the program is set at ten correct in two minutes for each of six consecutive days. Once Roy meets his aim, he will progress to the next step of this individually tailored receptive language program.

Should this plan fail to bring Roy to criterion performance within what Ms. Duncan feels to be a reasonable amount of time, she can make changes in one or a combination of the plan areas. Some of these options are listed below:

— *Program:* The time of day during which the program takes place might be changed, on the assumption that the child would perform differently either earlier or later in the day.
— *Programmed Event:* The teacher might try changing where she sits, how far away from Roy she places herself, or how she positions his head. She might also vary how she gives the cue, the intervals at which cues are given, or the words used to elicit response.
— *Behavior:* The length of the timing, how behavior is timed and counted, or the definition of correct and incorrect responses could all affect performance results.
— *Arrangement:* Changes in the number of events received for a correct response might make considerable difference in performance.
— *Arranged Event:* Another nutrient besides juice, or a change in event from the use of

FIGURE 7:16. Increase-Decrease Plan for Communication.

Plan Sheet Number 1. .... Location. Receptive Lang. Date 1-5-84.

Pupil Name Roy Jessup. Chronological Age. 8. .... Teacher Sheila Duncan.

Pinpointed by . Duncan.

PROGRAM	PROGRAMMED EVENT	INCREASE OR MONITOR BEHAVIOR	ARRANGE-MENT	ARRANGED EVENT	DECREASE OR MONITOR BEHAVIOR	ARRANGE-MENT	ARRANGED EVENT
1/5/84 10:00 to 10:15	T. seats Roy in his chair and takes seat in chair at 15" distance directly behind Roy	Turns head to cue: less than 5 seconds to respond = correct	Less than 5 seconds to respond : 2	Sip of orange juice from squeegee	Fails to turn to cue (e.g., moves head in wrong direction; or makes no turn)	More than 5 seconds to respond = error : 1	Ignore
Receptive Language: Turn to sound of voice		2 minutes		Praise			
M–F	Make certain Roy's head	Timed with stopwatch			More than 5 seconds to respond = error		
IEP Committee choice	is at midline; if not, position it there before giving cue	Counted with dual mechanical tally counter			Count on portion designated for errors on		
T. choice							

No. 1 objective: receptive language sequence Section: Turn to sound	Lean to one side of Roy's head maintaining 15″ distance and say into Roy's ear, "Roy, look at me." Give cue randomly from alternate sides at 10-second intervals.	T. charts on arithmetic chart Objective: 10 correct in 2 minutes for 6 consecutive days	dual mechanical tally counter T. charts on arithmetic chart

nutrients to some other type of reinforcer could have profound effects. Also, the use of praising and ignoring might be further examined.

Using an Increase-Decrease plan gives Ms. Duncan effectively twice the program-change options available in the two previous plan forms. She can concentrate her programming efforts on either correct or incorrect responses, or both.

An Increase-Decrease plan for managing a social behavior problem. Another teacher, Hazel Peters, uses the Increase-Decrease format to describe the plan she implements to decrease a particularly annoying social behavior problem in her Head Start classroom. One of Hazel's four-year-olds, Mattie, delights in rummaging through all the drawers in the room, spilling and spreading their contents at every opportunity. Ms. Peters decides that her first plan will simply be to count how often this problem occurs as she continues her current management strategies. She realizes that sometimes a behavior seems far worse than it is. Perhaps, for subjective reasons, the teacher finds a particular behavior impossible to endure, no matter how infrequently it occurs; or maybe the timing is such that the behavior seems more difficult to cope with than if it appeared during a different part of the day. Hazel feels that, before initiating further action, she needs an accurate picture of just how often Mattie engages in his search-and-destroy missions.

Figure 7:17 shows Hazel Peters' initial plan. According to the strategies outlined here, Hazel watches for occurrences of the pinpointed behavior throughout all the day's activities and counts what she defines as correct and incorrect behaviors on a grocery tally counter. Each time Mattie opens a drawer and closes it without dumping its contents, Hazel punches in a count on the "correct" side of the tally counter. At the same time, she praises Mattie. If Mattie cannot resist the urge to spill the drawer's contents, Ms. Peters reprimands Mattie by telling him he is not to engage in this behavior and to pick everything up off the floor. She enters the error count on the appropriate side of the tally device. Hazel sets an objective for correct opening and closing of drawers at 100 percent by October 16, one month from the initiation of this first plan. She'll take data on her current set of tactics for a week to see how things go. If no desired change seems imminent, Ms. Peters will write and implement a new plan, beginning the second week.

An Increase-Decrease plan to increase competence in oral reading. In a third, and final example, Richard Hanes describes the individualized reading programs he conducts as a part of his work in a resource classroom using the Increase-Decrease plan sheet.

One of the pupils who enters his room daily is eleven-year old Arnold Black, referred to the resource room as a result of his reading difficulties. Mr. Hanes performs a reading evaluation and places Arnold in material which he can supposedly handle, according to all the assessment information obtained. But he finds that Arnold spends much of his time fidgeting and staring off into space. His problem seems primarily one of concentration, and Hanes cannot attribute it to frustration with the reading material's difficulty level. It seems, rather, to be a matter of constant teacher attention. "Unless I can sit on him every minute, he just quits working." Even though the resource room has a much smaller pupil load than Arnold's regular classroom, Hanes doesn't have the time required to "sit on" Arnold, and besides, he feels it's a bad situation to encourage. Hanes is at his wit's end, until another resource teacher in the building suggests that he try using a tape recorder to keep Arnold involved. The plan Hanes devises is outlined in Figure 7:18.

Although Mr. Hanes works with a variety of reading programs and approaches, he places Arnold in the *Lippincott Basic Reading Series 2* because this series is the one Arnold's fifth-grade classmates are using. Arnold comes to the resource classroom for reading instruction every day between 10:45 and 11:45. According to the plan he devises, Mr. Hanes presents some initial instruction on other aspects of the reading program, and then places a tape recorder on Arnold's desk. He instructs Arnold to read into the recorder from a designated point in the reader, working independently until the teacher can return and contact him 15 minutes later.

FIGURE 7:17. *Increase-Decrease Plan for a Behavior Problem.*

Plan Sheet Number *1*.............. Location *Head Start Class* Date *9-7-85*.......

Pupil Name *Mattie A. Glover* Chronological Age. *4*...............

Pinpointed by *Peters*.....................

INCREASE OR MONITOR

PROGRAM	PROGRAMMED EVENT	BEHAVIOR	ARRANGE-MENT	ARRANGED EVENT
9/7/85 Classroom Complete school day	Scheduled activities throughout day in classroom (e.g., free play, snack, group instruction, story, music, recess)	Close drawer w/o emptying it T. counts w/grocery counter T. charts on arithmetic chart Objective: 100% by 10/16/84	1:1	T. praises Mattie after each successful handling of drawer

DECREASE OR MONITOR

BEHAVIOR	ARRANGE-MENT	ARRANGED EVENT
Dumps drawer T. counts w/grocery counter T. charts on arithmetic chart Objective: 0 drawers dumped each day by 10/16/84	1:1	T. reprimands (tells M. he is not to engage in this behavior and to pick up the spilled contents)

FIGURE 7:18. Increase-Decrease Plan for a Reading Program.

Plan Sheet Number 1 .... Location Resource Class Date 10-20-86

Pupil Name Arnold Black Chronological Age 11 Teacher Richard Hanes

Pinpointed by Hanes

PROGRAM	INCREASE OR MONITOR				DECREASE OR MONITOR		
	PROGRAMMED EVENT	BEHAVIOR	ARRANGEMENT	ARRANGED EVENT	BEHAVIOR	ARRANGEMENT	ARRANGED EVENT
10/20/82	T. places tape recorder on Arnold's desk	Say (read) words correctly	1 minute : 2	T. praise	Say (read) words in error (e.g., insertions, omissions, substitutions, miscalls)	1 error : 1	T. notes errors on sheet of paper
11:10 to 11:30	T. and Arnold find correct starting place in book. Briefly discuss section to read, what it concerns, new words coming up, etc.	T. counts with tally counter		T. gives feedback about correct			
M–F		T. times with stopwatch for 1 minute			T. writes each error on separate sheet	1 minute : 2	T. has Arnold read the error list
Lipp. 2							
T. choice: time, materials		T. charts on semilog chart			T. charts error rate on semilog chart		T. discusses any error patterns noted on list with Arnold
Objective: No. 5 in Master reading objectives list	T. tells Ar-	Objective: 100 words per minute					

—172—

with no more than 2 errors per minute for 5 consecu-tive days by 12/10

nold to read aloud into the recorder. In 15 min-utes, he'll contact him and to-gether they will listen to a por-tion of his reading.

T. contacts after 15 minutes when Ar-nold is reading aloud.

T. turns to random point on tape; T. and Arnold locate right spot in text and listen together.

Hanes contacts Arnold at the appropriate interval, making sure first that the boy is reading aloud at the time the contact takes place. He selects a random portion of taped reading and together he and Arnold find the place in the book corresponding to the recorded selection. The teacher then listens to a one-minute recorded sample on the tape, timing the sample with a stopwatch. While he and Arnold listen to the sample and follow along in the text, Hanes counts each correct and incorrect response using a tally counter, quickly noting error-words on a piece of paper. Errors are specified to include hesitations, substitutions, insertions, or omissions. (For some youngsters, Richard Hanes keeps precise data on each category of error, but he feels this isn't necessary with Arnold.)

At the end of the one-minute selection, Hanes stops the watch and the recorder. He gives Arnold feedback about the total correct and error count, which includes praise for his correct work. The teacher asks Arnold to read the list of error-words he has noted for the timing, and together they go over correct pronunciation and discuss whatever obvious error patterns might occur. Hanes charts the correct and error data immediately on a semi-log chart. The objective specified for this program is 100 words per minute correct, with no more than two errors per minute for five consecutive days by quarter review time, December 10th.

GETTING THE MOST OUT OF MEASUREMENT PLAN SHEETS

The plan sheets presented in this chapter are just one more step toward increasing and refining your classroom problem-solving skills. Used as part of a systematic approach to classroom measurement, they can be invaluable in helping you analyze difficult learning problems and achieve direction in the choice of successful teaching strategies and techniques.

Analyzing Instructional Problems

There are times when you find it almost impossible to put your finger on the "something" that's going wrong with an instructional plan. One of the most effective strategies we know of for dealing with a problem situation is to start by carefully describing the learning environment as it presently exists. While it's tempting to jump in with a variety of tactics to change the situation, our advice is: *Don't*. Before you go any further, try to isolate and describe the events currently surrounding the behavior of concern.

"No big deal," you say. "I can easily describe exactly what I'm doing." And yet, teachers are frequently surprised to discover that what they say they do and what they actually do in the classroom may be two very different things. The program envisioned while you sit at your desk or relax in your favorite armchair may vary considerably from the one implemented in the real world of your classroom.

Test how accurate your conception of a plan is by writing it out and taking the written plan into the classroom. Attempt to follow it as though you were a substitute teacher asked to fill in for the day. With no previous knowledge of the situation, could you follow the plan given? Better yet, get a colleague, volunteer, or aide to help you by trying the plan, using only the direction offered there. Don't be discouraged if first efforts require several rewrites before they actually describe the conditions you intend, and don't be content until you have written a plan describing the actual conditions with which you are starting.

Once you have identified the plan that is actually in effect, you are much better able to describe and assess the effects of any changes which might be introduced to this situation. You are also in a better positon to ensure consistency in presenting the environment you have outlined as necessary for a given child on a day to day basis.

Further, the process of simply trying to complete the plan and identify the necessary components can be immensely helpful in suggesting new ways to view a problem situation. Because each element may have an effect on performance, the way you are dealing with (and/or, might attempt to change) certain elements can now be reexamined in terms of their possible effects.

Using Measurement Plans to Their Best Advantage

A few hints will help you make the most of your measurement plans:

Be consistent. Once a plan looks the way you want it to, make sure it is implemented consistently. No matter how eager we are to carry out a plan, we may require some practice before we are sufficiently aware of our own behavior to realize whether or not we're following the plan correctly. Perhaps, for example, a child is supposed to be given feedback about correct problems every time he finishes a page. You or your aide may get so carried away with some other part of the program that you frequently forget this part of the plan. Initially, things go fine; but programs gradually lose their effectiveness when follow-through lapses.

If you think you notice a slight slip in plan implementation, sit back and analyze what's happening from time to time. Compare the plan in effect with the one you wrote. Are they the same? Make notes about inconsistencies directly on the plan; mull them over; talk them over with whoever is responsible for plan implementation if need be. Such checks often make the difference between a plan that is dynamite and one that seems effective one day, and fizzles out the next.

Introduce changes systematically. Using a systematic plan sheet usually means following the same plan for at least a three to five day period. When plan changes are made, they are generally initiated in only one area of the plan at a time, with a subsequent waiting period to see whether or not the change is the right one. Those of us eager to put each and every insight into immediate play may find initially that systematic planning runs counter to some of our deeper instincts. But teachers who implement changes on a "minute-by-minute" basis usually find that the bag of tricks empties out all too fast, and they've still to come up with a workable instructional program.

One discovery teachers make is that a plan sometimes needs time to grab hold and take effect. By allowing a trial period, you can be absolutely sure whether the plan you have chosen has any merit, may do in part or should be discarded. Often teachers who are initially resistant to this incubation period find that just watching the same plan in effect for a number of days gives them more ideas for what should be happening, and that these ideas are more accurate in the long run.

Implement a "standard" plan. You are likely to find that many of the plan sheets you write share several common features. To save effort in writing individual plan sheets, consider identifying these shared features on a master plan sheet, making several copies by the easiest means available. Leave blank spaces in relation to the key features of the master plan.

Figure 7:19 contains an example of this type of plan sheet. Note the blanks which can be filled out at a later time. Standard plan sheets have the advantage of identifying certain classroom routines and procedures that are always carried out in the same way. Pupils and other classroom helpers who have access to these plans find it easier to follow these procedures consistently when they are identified in a standard plan format. Some teachers report that it is easier to examine and develop individual variations in these routines for students who need them, once they have been spelled out in a standard plan.

Remember that measurement plan sheets are designed to last. We said it before, but we'll

FIGURE 7:19. A Standard Plan Sheet.

Plan Sheet Number . . . . Location . . . . Date . . . .

Pupil Name . . . . Chronological Age . . . . Teacher . . . .

Pinpointed by . . . .

PROGRAM	PROGRAMMED EVENT	INCREASE OR MONITOR			DECREASE OR MONITOR		
		BEHAVIOR	ARRANGE-MENT	ARRANGED EVENT	BEHAVIOR	ARRANGE-MENT	ARRANGED EVENT
SRA Reading Laboratory Science Research Associates	The teacher will present the pupil with the card in sequence for the day.	Say word correctly; — counts — records for — minute samples	1 handraise:1 —:— —:—	T. contact —points will be given for every — words.	Say word incorrectly	1:1	T. will provide correct word
Lab:	The pupil may read it over and raise his hand when he is ready for a one-minute sample.	Write letter correctly — corrects — counts — records	Contract: ___ ___ ___ ___	— points will be given for every — correct letter	Write letter incorrectly	1:2	1) T. will mark an X
Color:							2) Pupil will erase and correct response, then raise hand for recorrection. (Repeat this procedure until response
Date:				When the contract is satisfied, the pupil may:			
Time:	The pupil will then begin						

correct)

working on
the written
questions,
and raise
his hand
when he is
finished for
correction.

say it again. You don't have to rewrite the total plan sheet each time you change an element in your measurement plan. Many of the variables or events which we identify may be present and applicable over several days, weeks, or even months. For this reason, once you have carefully prepared an initial plan, you may find it necessary to make only minor alterations as the program progresses. These can be noted on the plan sheet with a date to indicate when the change was introduced.

CONCLUSION

The systematic measurement plan sheet has been presented as another of the problem solving tools you may choose to incorporate in your classroom. It should be clear that data collection can easily proceed without the use of such a plan, and there are many teachers who will elect to ignore this step altogether. But for those who are interested in using measurement plans, the benefits are many. The plan format provides a basic framework from which to analyze problem situations and examine the kinds of solutions you are presently attempting to apply. Teachers who have tried using the frameworks suggested in this chapter are often surprised and pleased at their increased problem solving ability in the face of the most difficult classroom situations.

As in all plan writing, it is true that these plans will involve some extra initial set-up time. Yet it is important to remember that dealing on a day-to-day basis with a problem situation which doesn't seem to be resolving itself not only wastes precious instructional time, but is a continuing source of anxiety and frustration to teacher and pupil alike. Since teachers must be master problem solvers, any tool which aides in this tremendous task is certainly worthy of consideration.

SECTION 3
CHARTING BEHAVIOR:
A PICTURE'S WORTH A THOUSAND WORDS

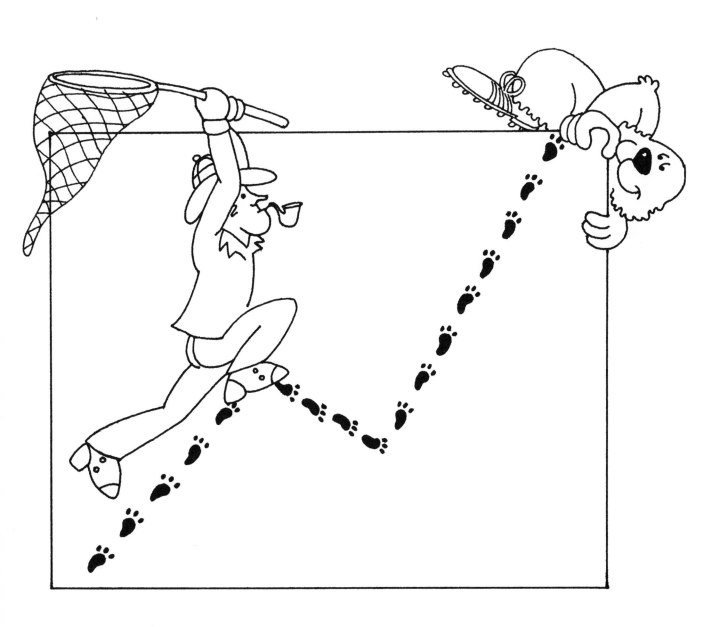

Introduction

Imagine yourself trying to explain to a friend how to drive cross country to your house. You've gone the route many times yourself, but what is the best way to direct someone who is unfamiliar with it? You could send him a collection of receipts for all the gas stations and hotels you'd visited between his place and yours. With time and a little luck, he might piece together all these fragments of information and deduce the fastest route. What a ridiculous idea, you say. Yet it's remarkably similar to the task you face each time you sit down to your grade book, miscellaneous notes, and anecdotal records to try to force these pieces into a picture of student progress. Calling your friend and describing the route in detail—a right onto Interstate 5; exit at the second turn-off after Lil's Tavern; two stoplights and then a left—is better, but still involves a lot of explaining on your part and possible misinterpretation on your friend's. The obvious solution is, of course, to draw a map.

In many ways, a chart is much like a map of pupil progress, tracking all the ups and downs, the starts and stops, along the path of learning. A grade book, or the other figures and notes you collect may tell the truth about performance; but a chart, like a road map, gives you the facts more interestingly, more impressively, and (most important) more clearly. You avoid the confusion and false conclusions that are so easy to fall into when you attempt to analyze long columns of figures to determine what relationship, if any, exists among them.

Charts are certainly nothing new to you. You find them tacked to the walls of your doctor's office, displayed in daily newspapers and prominently featured in weekly magazines. Initially, however, the idea of using charts in your classroom may seem somewhat strange, and even slightly intimidating. Rest assured that there is nothing difficult, frightening, or mysterious about charts. They are not the private territory of statisticians, nor need they be cloaked in heavy technical terminology. Charts require neither artistic skills to create, nor mathematical skills to interpret. A chart is simply a picture of a set of figures.

There are a number of charting forms to choose from. This section discusses five major chart variations, all of which can be adapted for displaying some kind of classroom data. Which chart you decide to use depends upon how often you take data, what performance measure you employ, and the way in which you intend to apply the data you obtain for making educational decisions.

Chapter 8
Using Pictographs, Pie Charts, and Bar Graphs to Summarize Data

Teachers often need a quick and easy way to summarize performance results, whether from a program implemented over an extended period of time, or from a one-time assessment of performance, such as that provided by an achievement test or diagnostic inventory. They want a vehicle for communicating summary information which is engaging enough to hold interest, and at the same time, simple and straightforward enough to be understood by all those interested in the child's progress, including the child himself. Pie charts, bar charts, and pictographs are designed to offer just such a summary of performance because they are simply constructed, easily understood, and visually appealing.

PICTOGRAPHS

Pictographs are exactly what their name implies: pictorial representations of important or interesting facts. They literally put into action the old proverb, "A picture is worth ten thousand words." At one time, pictographs were extremely popular in areas such as social studies, economics, and business, where they were commonly used to summarize raw score and percent data. For instance, a pictograph might be created to demonstrate the comparison of variables such as populations in different regions, or the production capabilities of competing companies. Recently, professionals in these areas have begun to favor more sophisticated graphing techniques, but pictographs are still frequently found in popular magazines such as *Time* and *Newsweek*, for making simple figure comparisons accessible to a wide range of readers. Pictographs can be useful in education, too, for presenting a simplified and interesting data summary.

Constructing Pictographs and Entering Data

There is no prescribed pictograph format. The "picture" part is usually a stylized drawing of an object which symbolizes the quantities being presented. Figure 8:1, for instance, contains a pictograph depicting the accomplishments of a group of fifth grade "Blue Ribbon Readers." Each ribbon represents one child's independent reading for the month of September, and the size of the ribbon pieces is taken to correspond roughly to the amount of extra reading completed.

FIGURE 8:1. *Blue Ribbon Readers.*

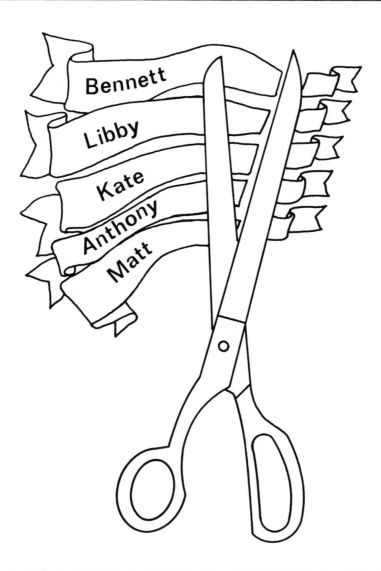

Another pictograph form, making possible more precise comparisons, is shown in Figure 8:2. This chart, too, relates information about extra reading, but adds more details by representing quantities of books read in terms of clearly specified units. These units take the form of small identical pictures of appropriate objects (in this case, books) shown in silhouette. Each pictured book represents two books read independently during the summer. Units could just have easily been assigned values of one, five, ten, or even more.

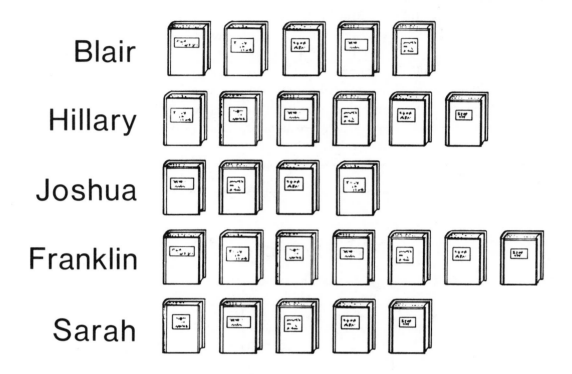

Summer Super Readers

Making the Most of Pictographs

While pictographs easily capture and hold attention, they also create some problems for the serious charter.

Overcoming chart imprecision. A pictograph such as that in Figure 8:1 often succeeds admirably in providing a gross overall impression of certain data relationships, but it resists precise interpretation. It is obvious that some Blue Ribbon Readers have done more than others, but how much more? Which of the ribbon dimensions is significant: the amount to the left; the tags to the right; a comparison of the pieces on left and right? And what about variations in ribbon *width*? Is amount related solely to length, or to total area, taking both length and width into account?

If you want to use a pictograph of this type, be sure to decide which single picture dimension is crucial in signifying comparative differences. For instance, let ribbon length be the important interpretive cue, while other dimensions, such as width, remain constant. Adding figures to the pictograph makes comparison even easier. For the clearest picture of performance, use a format similar to that in Figure 8:2, where calibrated symbols allow a more precise representation of performance amount.

Can pictographs show performance growth? Pictographs are summary charts, best adapted to showing a picture of overall or total performance as it appears at a specific point in time. The most serious problem these charts present for teachers is that they do not really lend themselves to a picture of performance growth: that is, behavior undergoing a continuing process of change.

The star chart is one form of pictograph which tries to incorporate the idea of growth. Stars are added to the chart each time the child completes a desired behavior, bringing him one step closer to a specified goal or objective. For years classroom bulletin boards and kitchen refrigerators have blazed with star charts, rockets-to-the-moon, mountain climbing expeditions, and similar works of pictographic fantasy. While they permit enthusiastic astronomers and astronauts to mark their progress in a creative and colorful fashion, such charts are not designed to give teachers the kind of precise, immediately accessible and easily interpreted information needed for ongoing educational decision making. Figure 8:3 shows a star chart used by an enterprising fourth grade teacher to reward her youngsters for "helpful" behavior at various periods (e.g., recess, lunch) throughout the day. The teacher counts the stars at the end of the week and presents citizenship awards to those youngsters who have enough accumulated. Her chart helps her *summarize* the total week's performance, but is not very helpful if she wants a quick assessment of behavior change from day to day.

FIGURE 8:3. *Star Chart*.

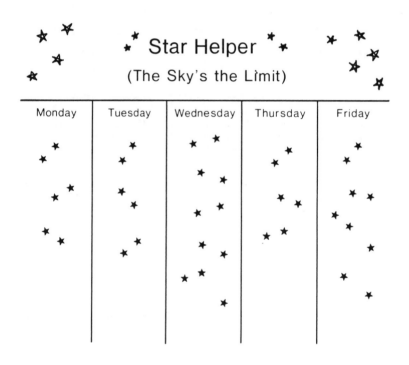

PIE CHARTS

The pie chart, also known as a *circular percentage diagram*, is a circle divided into a number of pie-shaped wedges which demonstrate how various portions of a total are related to one another. Pie charts are often used to summarize percent data, comparing related pieces of information, each of which represents a percentage of the whole. The simplicity of pie charts, both to make and to read, has real drawing power. Since our currency system is based on decimals, the pie chart, which expresses amounts in terms of hundredths, relates directly to that system. For this reason, circular percentage charts are a favorite with energy czars, presidential candidates, and advertisers who are trying to put the pinch on your dollar.

To the teacher who needs to summarize and compare performance data, the simplicity and ease of communication this charting form promises also has considerable appeal.

Constructing Pie Charts and Entering Data

The pie chart derives its name from more than just its circular shape. Like a pie, the chart is divided into a number of wedge-shaped components, or slices, formed by lines extending outward from the center of the pie, and intersecting the pie circumference at different points. These slices are called sectors, each sector representing a certain proportion or percentage of the total pie. The larger the percentage, the larger the sector. Following certain charting conventions makes your pie chart easier to create and interpret.

Determining sector size. Normally a circle's circumference is considered to be divided into 360 equal invisible segments called degrees. But since a pie chart represents percentages, it is a special kind of circle, whose circumference is divided into 100, rather than 360, equal segments, each representing one percent (Figure 8:4). A pie chart could theoretically contain 100 sectors, each worth one percent of the pie. Normally, however, pie sectors are much larger than this. For instance, a wedge representing a score of 25 percent would take up exactly one quarter of the pie; 50 percent exactly one-half; 75 percent, three quarters; and so on.

Arranging pie chart sectors. To make pie charts consistent and easier to interpret, sectors should be arranged according to size, from largest to smallest, proceeding clockwise around the circle. The total number of sectors depends upon the data you are presenting.

Suppose you decide to create a pie chart that will show the school psychologist why a particular child in your classroom is making your life next to impossible. You call the youngster Fast Fanny—every time you turn your back for a second, like a shot she's out of her seat, out of the door, and down the hall. Not only that, but most of the time she's in her seat, Fanny is daydreaming, not really attending to her work.

One week you collect some data, hoping to find out exactly how much of her time Fanny spends *on task*, compared to *out-of-seats*, and to *daydreaming* (a chunk which you have carefully defined). This will be a classic percent-of-time exercise. Simply time *out-of-seats* every time they occur, starting a stopwatch the instant Fanny's fanny leaves her chair without permission, and stopping the watch when she returns to her desk. *Daydreaming* is timed in the same fashion. *On task* can either be timed in the same way, or derived by subtracting combined *daydreaming* and *out-of-seat* time from overall seatwork time allowed. At the end of the week, your raw data summary looks like this:

Out-of-seats	*60% of time*
Daydreaming	*30% of time*
On task	*10% of time*

When these data are entered on a pie chart, it looks like the one in Figure 8:5.

Making the display dramatic. For greater clarity, different sectors are often distinguished by coloring them in bright, contrasting colors, or bold patterns and shadings. Each sector should be labeled, and percentages or figures may then be entered directly on the appropriate sectors. Fanny's data, for instance, have much greater impact when the segments are shaded and labeled as in Figure 8:6. The segments yield a bold, immediate visual representation of the proportional differences in Fanny's on-task and inappropriate behavior.

FIGURE 8.4. *Pie Chart's 100 Segments.*

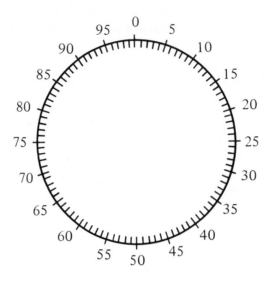

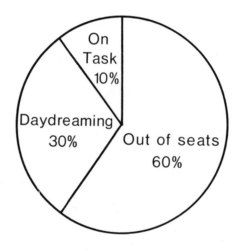

Making the Most of Pie Charts

Before deciding to adopt pie charts as a part of your own classroom decision-making appa-ratus, be sure that you understand some of the restrictions this charting form involves, as well as how the chart is best employed.

Keep the number of pie segments reasonable. If your data are extremely complex, using a pie chart to display them may be more confusing than helpful. In Figure 8:7, a teacher has summa-

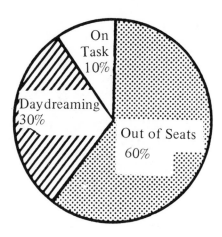

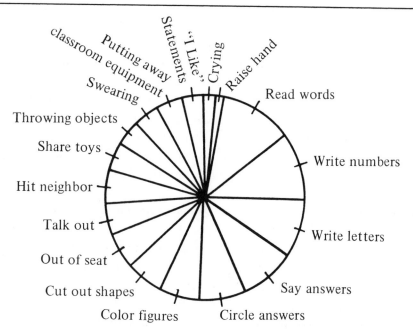

rized duration data, showing the average amount of time a particular youngster has spent during a given week engaged in each of a number of behaviors identified as critical. Obviously, there are too many pieces of data here to gain any but the most blurred impression. Remember, pie charts are best put to use to show a limited amount of data.

Remember that pie charts used to display percent can hide certain issues. Percentages can fool you. You'll remember from the discussion of this measure in Chapter 5, that even though a child's percent of errors may be decreasing, if overall responses attempted are increasing, the total errors will be the same or even greater. This trend may be concealed by the present measure, and

A Day at the Races

the confusion compounded when data are translated onto the reporting form.

Take the example of Maxwell Moneybags, who is trying to reform the spending habits of his wife, Mennypenny. Maxwell finally requires that Mennypenny account for her accounts. He asks her to show him a pie chart demonstrating where the funds have gone at the end of each month. But Penny pulls a fast one. She shows Max a pie chart which tells him what percent of the amount she spends falls within each budget category. She does not, however, designate the total (raw score) on which the percentages are based. While the chart (Figure 8:8) shows that Menny's afternoons at the track are using up a decreasing proportion of total expenses, the raw amount upon which percentages are based might be actually growing each month. Mennypenny could be milking Max for millions!

Problems in trying to show performance growth and change. Like their cousins, pictographs, pie charts are static charting forms. They are best at presenting data summaries, where data are generally added together and compared to a whole. They are not good for showing performance change, which requires documenting multiple comparisons of data taken at a number of different times.

Suppose you decide to use a pie chart to keep track of a pupil's progress in a buttoning program. Each day you count the number of buttons the child buttons correctly and incorrectly, then calculate the error percentages. At the end of the week, your data are as follows:

	CORRECT	*ERROR*
Monday	*0%*	*100%*
Tuesday	*20%*	*80%*
Wednesday	*35%*	*65%*
Thursday	*50%*	*50%*
Friday	*75%*	*25%*

Summarizing all the data, by adding and averaging and then putting them on a single pie chart, you would get a picture like the one in Figure 8:9. The chart displays overall errors for the week which are approximately twice as many as correct responses: not very impressive. But what about the

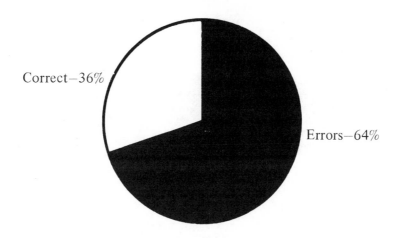

changes that this pupil's behavior has undergone on a day-to-day basis? From zero percent on day 1, to 75 percent correct on day 5! The pie chart fails to show you a crucial performance aspect—*growth*.

You could, of course, draw a new chart each day (Figure 8:10). But as you can clearly see, even with only five days' performance displayed, your charting task would be awkward and the data soon unwieldy. Think of charting, organizing, displaying, and storing a new pie chart for each day of a child's performance over a 30-day period. To say nothing of comparing, interpreting—digesting, if you will—all those pies.

The pie chart at its best. Although they are not well suited to the display and analysis of discrete daily performance changes, a well thought out pie chart can present an accurate and easily interpreted picture of gross, overall program constituents.

One teacher in a special education classroom uses pie charts to summarize the programming emphasis adopted for each of her pupils. At the beginning of each month, the teacher creates pie charts summarizing the percentage of time per day allotted for programming in each of eight developmental areas. Because instruction is individualized, the amount of attention devoted to each of these areas is, of course, different from one child to another. The teacher realizes how easy it is to concentrate so wholeheartedly on an area in which a child expresses marked deficits, that other areas are slighted or totally ignored.

From month to month the charts inform her exactly how total program emphasis changes as discrete programming decisions are made. In the charts she draws up for one child in the months of February, March, and April (Figure 8:11), it is possible to see a gradual increase of program emphasis devoted to reading, until that area is found to take up 60 percent of program time. At this point, the teacher carefully assesses whether other areas are being given the recognition they deserve. Based upon her analysis, she may juggle the child's scheduling for the next month, so that other vital areas receive more consideration. In this fashion, a more balanced overall plan is achieved.

Another teacher uses a pie chart (Figure 8:12) to summarize data he has collected concerning a new youngster's reinforcement choices during the first weeks of placement in his classroom. Each time the child earns free time, the teacher notes what general toy or activity choices are made (e.g., art, music, blocks and construction, picture books). At the conclusion of the day, the teacher tallies

FIGURE 8.10. *Problems Depicting Performance Day to Day.*

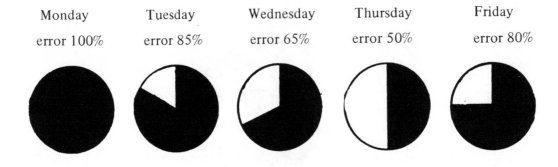

Monday	Tuesday	Wednesday	Thursday	Friday
error 100%	error 85%	error 65%	error 50%	error 80%

FIGURE 8:11. *Pie Chart Summarizes Shifting Program Emphasis.*

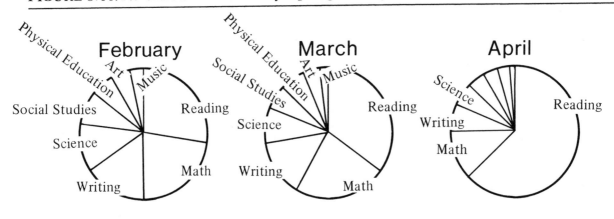

FIGURE 8:12. *Pie Chart Summarizes Reinforcement Choices.*

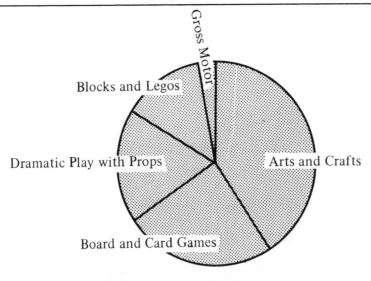

choices under each of the activity categories available, and adds all tallies together over the two week initial placement period.

Notice that this particular pie chart is created to handle raw score rather than percent data. Since the child made a total of 80 choices during the assessment period, the total chart represents a raw count of 80. Half the chart represents 40, and so on. Reviewing the data as they are displayed on the pie chart, the teacher is able to see clearly where the child's major areas of interest are. He can now "beef up" these activity areas, giving the child numerous new selections in favorite categories, thus ensuring the success of free time as a reinforcer for academic performance. In some cases, the teacher may add what he feels are enticing choices to the categories the youngster seems to ignore, hoping that by increasing their drawing power, he will encourage the child to try new reinforcer categories, expanding the youngster's range of interests and capabilities.

BAR GRAPHS

The bar graph is the simplest of all charting forms, consisting of a straight line drawn wide enough to be easily visible, whose length represents the size or amount of a specified entity (e.g., behavior). Although bar charts have much more flexibility than pie charts, like pie charts they are best suited to presenting data summaries. Bar graphs are excellent for providing a "performance profile" on a one-shot basis: for instance, summarizing the results of a standardized skill inventory or achievement test. The height of the bars can be compared for a very simple, very broad picture of performance. Three distinct bar chart formats (plain, component, and compound) give teachers a wide range of options with which to display classroom data.

Basics in Bar Graph Construction

Because the bar chart is such a popular form of data display, you'll find charting paper easily obtained at numerous office supply firms. Whether you buy chart paper or construct your own freehand charts, however, certain basics are still up to you. These include determining the chart *range*; selecting the charting *scale*; and deciding which of the three chart *varieties* best represents your data problem.

Determining the range. This simply means stipulating the upper and lower limits of the data which will be displayed on the chart. Convention generally dictates that zero is at the bottom of the bar chart scale. The rule of thumb for establishing the range's upper limit is to estimate the largest possible figure to be charted and round it off slightly upward to assure ease of scale computation.

The actual size of the upper data figure depends upon a number of factors, including limits imposed by the nature of the task itself and by the measure of performance being used. For instance, a teacher who consistently assigns a page of math problems with 50 items on the page would probably use 50 as the upper limit of the chart. If the number of problems assigned varied from day to day, the top of the range would include the largest number ever assigned. Certain measures also affect the data range. If a percent measure is used, the range will, of course, always be zero to 100. Charting data in terms of a measurement set-up such as trials to criterion, on the other hand, might require a range of zero to ten, assuming ten is the maximum number of trials allowed.

Once you have decided what the upper and lower limits of the data are to be, record these limits at the top and bottom of the amount scale, located at the left of the chart.

Assigning vertical amount scale values. The amount scale and its divisions are generally designated next to the vertical line to the left side of the chart. (You may choose to make a horizontal bar chart, although vertical is the norm.) If you are drawing your own chart, the number of major scale divisions is fairly flexible. If you are using commercially available paper, you'll usually find that the scale has major interval divisions already marked on it and your job is simply to assign appropriate numbers to the interval markings.

For the sake of convenience and uniformity, most chart paper designed for business purposes has been assigned 100 small divisions, and ten larger intervals marked along the vertical amount scale. Such paper is perfect for constructing percent bar graphs. Numbers may be reassigned, however, according to the individual scale necessary for your data. Suppose, for example, that you are keeping raw score behavior counts on a particular behavior which you have observed to go as high as 200 instances in a single day. Your range would be zero to 200, and scale divisions might be labeled in increments of 20 (i.e., 0, 20, 40, 60, 80 and so on), until the top of the range is reached. Data ranging from zero to 50, on the other hand, might be charted on a graph whose vertical scale divisions increased in increments of 5 (i.e., 0, 5, 10, 15, 20, etc.).

Major scale divisions are frequently evenly divisible by units of 5 or 10 for ease of figuring, making estimation of values between major scale divisions quick and simple. Make sure that the values are neither so large that you lose sensitivity to small changes, nor so small that construction becomes a cumbersome ordeal.

Choosing the Right Bar Type

Once you have established the chart's range and scale, you are ready to enter data on it by constructing the bar portion of the graph. Your bar can be as narrow, as wide, as simple, or as complicated as you wish. Just remember that the critical dimension is bar height, and try not to mask that dimension with a lot of unnecessary fancy footwork. Using basically the same kind of format (a labeled vertical amount scale running down the left side of the paper, meeting a horizontal record floor that extends from left to right at the bottom of the chart) you can create a variety of chart types, each relating data according to a slightly different perspective.

The plain bar. The simplest bar format is the plain bar chart. Such a chart contains one or more bars, each of which represents the straight amount, size, or quantity of a specified entity (e.g., behavior) at a particular time. Figure 8:13 illustrates one way in which a plain bar chart may be constructed and used. Here test results are recorded for each of nine children screened on a math inventory administered at the beginning of the school year. Each plain bar represents the performance of a single child. The amount scale is calibrated to show percent correct.

The component bar. The component bar graph is composed of a single bar, divided into a number of components each of which represents a single part of a designated whole. Component bars are especially popular for charting percent data, drawn so that the entire bar is taken to represent 100 percent; and each component is proportionate in size to the percent of the whole for which it stands. In Figure 8:14, a component bar displays data collected on several smaller pinpoints comprising a larger behavior chunk. The teacher has used the bar to analyze duration data; specifically, what percent of time a particular youngster in her preschool classroom spends engaged in each of three different types of social interaction (i.e., cooperative play, parallel play, and isolate play). The bar components have been arranged in an ascending order, beginning with the largest component (percent of isolate play) at the bottom. Different shadings or cross-hatchings make component differences clearer, the darkest shading at the bottom of the chart. The chart is keyed, indicating what each component represents, with a word on or beside the bar.

The compound bar. Used less frequently than plain and component bars, the compound bar is constructed by joining two or more bars laterally to show a data relationship or comparison. Each bar is marked to distinguish it easily from its neighbor. An example of this charting form is found in Figure 8:15, where a teacher has used compound bars to compare the performance of three children in reading achievement tests given at three separate times during the school year.

FIGURE 8:13. *A Plain Bar Chart.*

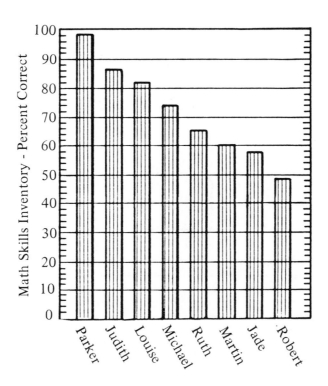

Making the Most of Bar Charts

Consider carefully both the potential and the limitations of bar charts before putting yourself unreservedly behind bars.

Keep data comparisons simple. When a bar chart is used to compare too many variables, the result is a confusing picture that is very difficult to interpret.

The bar chart format is handy because of its simplicity. Forcing it into a complicated function it is not intended for, defeats its basic purpose.

Use the bar chart as a summary display form, not to portray growth or change. Don't rely on a bar chart to provide any kind of clear picture of minute, frequent performance changes. Although bar charts are sometimes used to display data taken at frequent intervals (e.g., three times a week or more), this is a misuse of the bar format.

FIGURE 8:14. *A Component Bar Chart.*

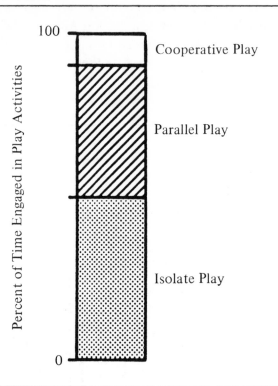

FIGURE 8:15. *A Compound Bar Chart.*

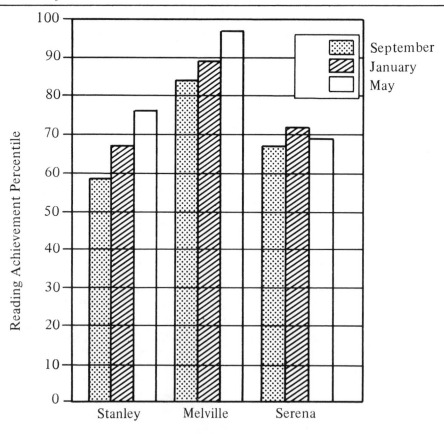

The chart in Figure 8:16 makes clear some of the obvious drawbacks in using a bar chart to display and analyze data taken on a frequent basis. This chart shows the percent of recess time a preschool youngster spends each day engaged in active play. The child's teachers and parents were concerned with his hesitation to participate in vigorous play activities: He always stood quietly by and watched while other children rode trikes, climbed ladders, slid down slides, and careened around on wagons. After a week of initial data collection, the teachers initiated a plan hoping for success in changing the boy's level of active play.

The figure demonstrates some problems involved in the bar chart form. The combination of continuous, frequent data and lateral joining of the plain bars results in an unattractive data "blob." Although an immediate performance improvement is apparent, what happens after the initial change is not so easily deciphered. Bars tend to become confusing if numerous, and especially so when they record alternate increases and decreases, as in this example. It is much easier for the eye to follow a single line across the face of the chart (as you will see in the next chapter) than to jump from bar top to bar top. Being able to draw such a line on a chart allows you to analyze both how the child has been doing, as well as how he can be predicted to do in the future. And, since the dual potential for data analysis and prediction are what make daily data collection invaluable, another chart form is obviously required in this context.

FIGURE 8.16. *Problems Showing Performance Change.*

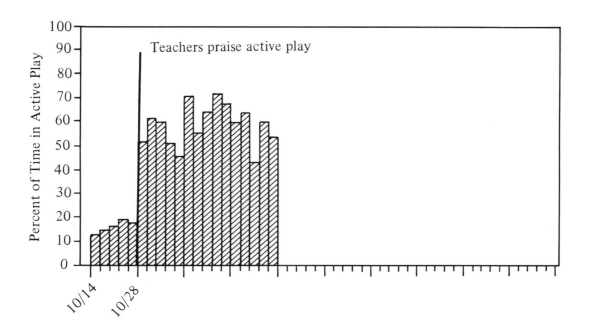

CONCLUSION

The charts described in this chapter are in a very basic sense *static* in the picture of performance they present. They are designed to show how the child is functioning at a specific point in time or to give a condensed summary of his performance over a longer period. There are times when such a summary might be valuable to you. If, however, your bigger concern is behavior change on a more frequent and immediate basis, you will need a very different charting form.

Chapter 9
Line Charts: A Dynamic Picture of Performance

Just as children never stay in one place for long, neither does their behavior. Movement, growth, and change are elements basic to the nature of classroom performance. If a program is working well, if the child is operating successfully within his learning environment, that change is positive. The child does better and better. He may perform a specified behavior more frequently, for longer periods of time, or with greater accuracy. If the youngster is having difficulty, usually the opposite is true. Capturing performance changes accurately and with enough regularity to make timely program alterations requires a chart which provides a moving, dynamic picture of performance.

The line chart is better suited than any other to presenting just such a dynamic picture of data fluctuation and change. The chart gets its name from the fact that data entries are connected to form a line which moves across the chart from left to right. Following the dips and peaks this line takes as it traverses the chart face tells you at a glance how performance stands in relation to the last time it was recorded, and whether or not the behavior being measured is on the desired course.

Two basic varieties of the line chart have been developed and refined over the years for use in business, economics, and in the natural and social sciences. Recently these charts have been appearing more and more frequently in the classroom to describe the results of that combination of science and art known as teaching.

THE ARITHMETIC CHART

Like all line charts, the arithmetic chart (or linear graph) is designed to show the changing magnitude of a single variable (e.g., behavior) over time. The chart generally takes a standard format, with the vertical axis used to indicate the level of behavior, and the horizontal axis used to depict the passage of time. Figure 9:1 contains a variety of arithmetic chart forms designed by teachers for

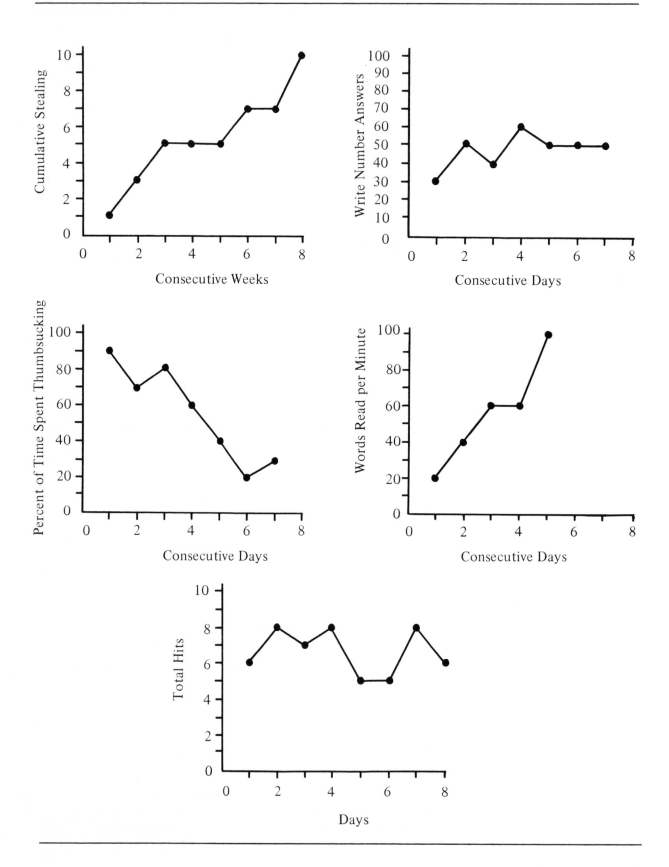

classroom use. Each chart illustrates a slightly different combination of behavior and time variables. Notice that the vertical axis refers in each case to amount of behavior, although behavior may be defined in many different ways. The horizontal axis describes the time frame in which observation or assessment takes place.

The arithmetic chart is often called an *equal interval* chart. This is because each interval along the amount scale of a particular chart is of exactly the same size and value as every other interval along that scale. The chart's equal interval construction gives us a very special picture of performance: Equal distances on the chart mean an equal amount of change. A specific amount of improvement, therefore, will look the same no matter where along the amount scale that improvement takes place. On the arithmetic chart in Figure 9:2, for instance, an increase of ten looks comparable, whether the child is working on a skill which appears to be new and difficult for him (e.g., increasing performance from 0 to 10), or adding slightly to a skill which is apparently well established (e.g., going from a score of 100 to 110). This view of performance growth is in distinct contrast to that displayed by the semilog chart discussed later.

Arithmetic charts offer many advantages, including relatively low cost, simple construction, rapid charting, and ease of interpretation. They are extremely flexible: Using an equal interval chart, you can record data for each and every one of the measures presented in this book. Most important, the arithmetic chart is the first we have discussed so far that is specifically designed to look at behavior change on a frequent basis. Data can be plotted on an arithmetic chart however frequently samplings occur, whether monthly, weekly, daily, or even more frequently (e.g., recording behavior at each of several sessions during a day, and charting each session separately).

FIGURE 9:2. *The Arithmetic Chart: An Equal Interval Chart.*

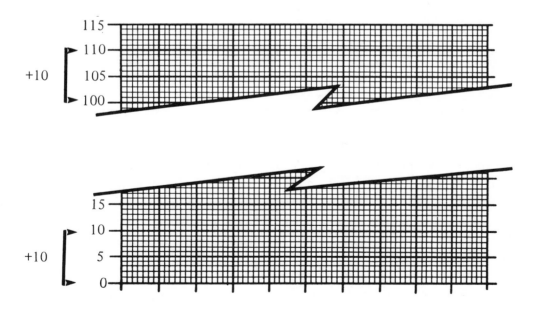

While it is often possible to purchase a commercially made arithmetic chart which suits the particular charting job you have in mind, frequently you will find that you must buy chart paper and construct the body of the chart—the vertical and horizontal scales—yourself. If you do not have access to ready-made charts, the following tips may be helpful.

Charting paper. Arithmetic charts are generally constructed on quadrarule (square rule) paper, so called because it is comprised of numerous small, finely printed squares contained within larger, more heavily ruled squares (Figure 9:3). The bigger squares are called *macros*, and the smaller, *micros*.

The number of micros per macro may vary considerably, depending upon the type of paper you purchase. Regardless of the micro/macro ratio, these squares form a ready-made grid, upon which the arithmetic chart is easily constructed. The heavy macro lines assist in identifying major intervals along both horizontal and vertical axes, and the finer micro lines aid in locating data points which fall between those intervals.

FIGURE 9:3. *Macros and Micros.*

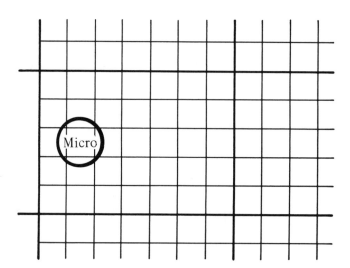

Constructing the vertical amount scale. Although there's no hard-and-fast rule, the amount scale generally runs vertically up and down the left side of the chart. Commercial chart paper may carry a ready-made vertical scale, or may leave the numbering up to you. To construct your own vertical scale requires taking into account such factors as data range, the macro/micro distribution of the paper you happen to be using, and the number of major and minor scale intervals you feel will be most convenient and sensitive to your data.

The range on an arithmetic chart, just as on a bar chart, begins at zero and extends to include the upper limits of the data. When measures such as raw score or rate are being charted, the upper limit of the range is established by rounding off slightly upward of the highest likely figure. Measures such as percent involve an automatic performance ceiling (i.e., 100) which becomes the top value on the amount scale.

Major interval values are assigned to the macro marks along the left side of the chart. These values should be of an appropriate size in relation to the scale's range. In general, the larger the range, the larger the value assigned major intervals (e.g., For a very large range, say zero to 1000 or greater, major intervals of 10 would be too small, resulting in a scale that was extremely long and unwieldy.). To make calculation easier, values are generally divisible by 2, 5, 10, or multiples thereof. Remember, too, that major division values must be consistent with minor divisions. For example, major intervals designated 0, 5, 10, etc., should not have two or four minor intervals, since each minor interval would then have to be assigned cumbersome values such as 1.25, 2.5, 3.75, etc. Figure 9:4 contains a vertical scale whose range is 0 to 100, with the value of each macro increasing by ten, and minor intervals worth two units each.

As you construct the vertical scale, keep in mind that arithmetic charts are *equal interval* charts. All major divisions on the scale are an equal distance apart and of equal value. The same holds true for the relationship of minor scale divisions.

FIGURE 9:4. *Vertical Amount Scale on Arithmetic Chart.*

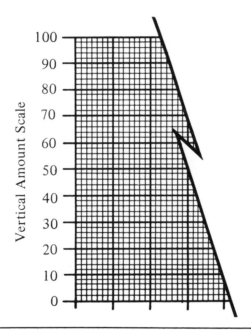

Constructing the horizontal time scale. The line running along the bottom of the chart is the horizontal time scale. This line, by convention, is usually drawn slightly longer than that for the vertical scale for visual contrast. Like the vertical scale, it should be far enough away from the border of the chart paper to allow for easy, clear labeling.

The time scale may be set up in a variety of ways, depending upon how behavior is sampled. If data are collected on a per-session basis, for instance, the results of two or even more samples per day might be recorded, each sample entered on its own individual line. For data collected and recorded weekly (i.e., the results of a quiz or test), each chart interval represents one week. Classroom data are frequently collected on a daily basis, especially if the data analysis techniques presented in Chapter 11 are to be applied. For daily data, the horizontal time scale might look like the one fea-

FIGURE 9:5. *Horizontal Time Scale on Arithmetic Chart.*

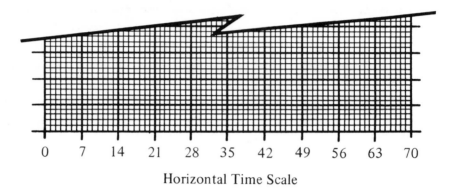

0   7   14   21   28   35   42   49   56   63   70

Horizontal Time Scale

tured in Figure 9:5. This scale is constructed on a successive calendar days basis, with each interval mark representing a specific day of the week. Monday's data would be recorded on the first day, Tuesday's on the second, and so forth. In this particular example, Saturday and Sunday are also accorded intervals on the time scale. Including Saturday and Sunday lines increases the uses to which a chart may be put, making it possible to chart data taken by parents over a weekend, if such data should be desired. The heavy macro lines on the time scale represent Sundays, and help the charter identify where each school week's new data cycle begins. An alternative successive days system, with macros every five units, might omit weekends entirely. Whatever your basis for constructing the horizontal scale, decide what system you will use before setting up the chart, and be consistent as data are entered.

As on the vertical scale, divisions along the horizontal time line must be equidistant from one another, and each scale interval must be equal in value to every other interval along the scale. The scale's proportions need not, however, be identical to those of the vertical amount scale. Just remember that the relationship of interval size on the vertical scale to interval size on the horizontal is crucial to the shape of the resulting performance picture. This is evident in Figure 9:6, where the same data are charted on three different grids. As you can clearly see, if amount scale units are much smaller than those on the time scale, the data curve appears to flatten out and wander aimlessly across the surface of the chart (Figure 9:6a). When these proportions are reversed (Figure 9:6b), the curve takes a steep upward climb, appearing ready to shoot through the roof of the chart. In Figure 9:6c, amount units and time units are equal in size, and the resulting curve pursues a more orderly course, presenting a more representative and, on the whole, more easily interpreted picture of what is happening. As you set up your own arithmetic charts, consider carefully the influence that grid spacing will have on your data. Try to visualize how performance patterns you might expect to see will appear on the chart you have constructed. Get to know the chart you have selected before you put it to work in the classroom.

FIGURE 9:6a. *Vertical Scale Increments Smaller.*

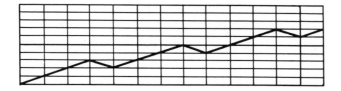

FIGURE 9:6b. *Vertical Scale Increments Larger.*

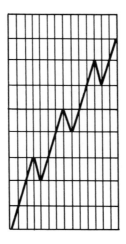

FIGURE 9:6c. *Amount and Time Units Same Size.*

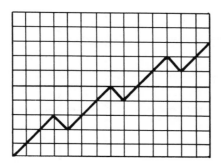

Entering Data on the Arithmetic Chart

The more frequent your data samples, the more frequent are your opportunities to use performance as a guide in decision making. If you collect data on a daily basis, it makes sense to chart the data each day. That way, you have a chance to intervene immediately if performance trends appearing on the chart indicate that a program change is necessary. The steps required to enter data on the chart are almost as easy as 1-2-3.

Step 1: Figure the data. Perform on the raw data whatever calculations are required by the chosen behavior measure before entering the data on the chart. The scores in the example in Figure 9:7 represent percent correct obtained by a child on a series of daily arithmetic work sheets. Percent, remember, is arrived at by dividing the total number of responses into the number of correct or incorrect responses (in this case, just correct).

Step 2: Locate the right amount interval. Working upward from zero, find the vertical scale interval that represents the amount (whether raw score, rate, percent, etc.) you wish to record. For instance, your percent chart may look like the one in Figure 9:8a, on which major scale intervals are marked off in units of 10. Since the first day's score is 44 percent, find the 40-percent mark, and then continue upward to the minor scale interval representing 44. Since each minor interval on this chart is worth two percentage points, count two lines up from 40 percent to arrive at the appropriate interval. The data plot will not always fall directly on an interval point. If the score were one like 45 percent or 47 percent, it would fall halfway between two minor interval lines.

Step 3: Locate the right time scale interval. The record sheet shows that the first day's data were collected on a Monday. Find the appropriate Monday line on the chart and follow it up until it crosses the appropriate amount line (Figure 9:8b).

Step 4: Make the data plot. At the point where amount and time lines intersect, a data plot is made (Figure 9:8c). In this case, correct responses are symbolized with a small dot.

Step 5: Connect the plots. Data points are connected sequentially from left to right across the chart to form a line whose upward and downward progress describes changes in pupil performance over time. In Figure 9:8d, two weeks of data points are plotted and connected. Saturday and Sunday lines are left blank, since no data were collected on these days. The break between Friday's and Monday's data is a natural way to indicate an instructional break which may affect subsequent performance.

FIGURE 9:7. *Sample Raw Data: Percent Correct.*

Weeks	Mon	Tues	Wed	Thurs	Fri
1	44	40	46	48	48
2	46	50	52	48	50
3					

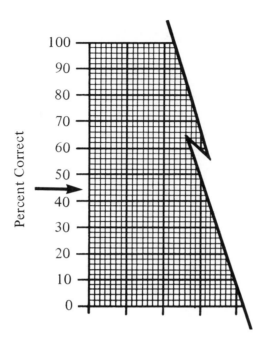

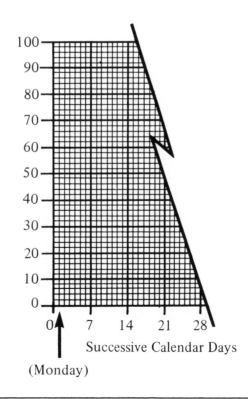

FIGURE 9:8c. *Entering Data: Making the Data Plot.*

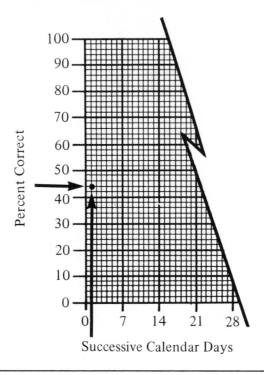

FIGURE 9:8d. *Entering Data: Connecting the Plots.*

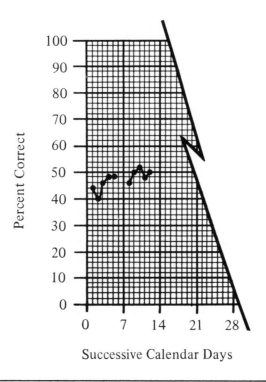

THE SEMILOG CHART

The semilog chart is a graph that looks funny, but works magic. In contrast to the arithmetic chart, it features an amount scale based upon logarithmic rather than linear increase.

In all likelihood, it's been years since you were on speaking terms with a logarithm, and maybe you're in no hurry to renew the acquaintance. If you think back to your last algebra class, you'll remember that you multiply by adding logarithms and divide by subtracting them. Whatever you do or don't remember about logarithms, you should have no difficulties with them when it comes to a semilog chart. All the mathematics are taken care of when the chart paper is printed. The intervals along the vertical amount scale are constructed in such a way that they're spaced in accordance with the logarithms of the numbers you see assigned to these lines. That's why these intervals look so strange. But the chart divisions carry natural numbers, making the process of entering data on it exactly the same as with any other line graph. In other words, you can reap all the benefits this particular chart has to offer, without any extra effort or special math background.

A Different Way of Looking at Performance

The semilog chart differs dramatically from the arithmetic chart in the way it portrays performance growth and change.

Emphasis on proportionate nature of growth. You'll remember that the arithmetic chart describes behavior increase or decrease in terms of amount, with equal distances on the chart showing equal amounts of change. The semilog chart shows behavior increase or decrease in terms of percent of gain or loss, with *equal distances showing equal percent of change*. This difference is easily illustrated by the charts in Figure 9:9. On the arithmetic chart (Figure 9:9a), the performance of Virginia, who gets 50 problems correct one day and 100 correct the next, is compared to that of Vanessa, who gets two correct the first day and four correct the following. Because this is an arithmetic graph, the difference in the performance of these two youngsters appears staggering. The arithmetic chart represents amount of change as additive and absolute: A gain of 50 in absolute terms is much larger than a gain of two.

The semilog chart (Figure 9:9b), on the other hand, shows the size of the performance gains of each as being the same. This results from the fact that the *percent of change* for each is exactly the same. This can be demonstrated by figuring percent of change for both girls. Take Virginia's scores first. She adds 50 correct to a previous score of 50. Percent of change is figured by dividing the amount of increase by the previous score:

50 divided by 50 = 1.00 or 100% change

Vanessa adds 2 correct to a previous score of 2. When her percent of change is figured:

2 divided by 2 = 1.00 or 100% change

The percent of change is exactly the same.

Individuals in business and economics have been quick to recognize the advantages of the semilog chart for calculating and representing variables such as growth rate or compound interest, where percent or ratio of change is the most important aspect. In recent years, certain educators,

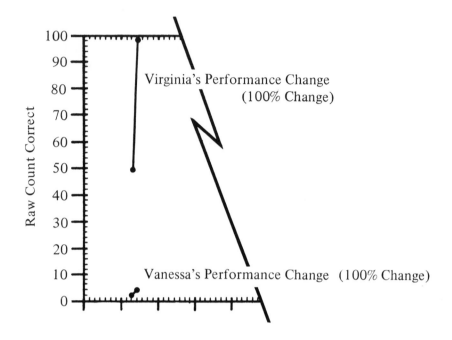

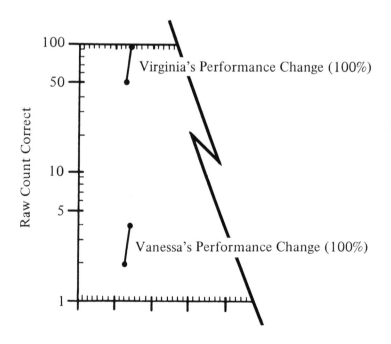

too, have shown interest in the chart because they agreed with its description of learning in terms of proportionate change. They feel, for instance, that anyone watching a baby learning to take his first steps would agree that the process could not totally be described in terms of additive, or absolute, change. Those first steps are a quantum leap for the infant, with the steps that come after, simply refinements.

These educators maintain that most growth is proportional in the same way: The efforts that get us to "learning" are of much greater difficulty, and hence, perhaps weighted differently, than those that merely represent practice, refinement, and strengthening of a learned skill. For the kindergartner just learning to read his first words, a jump in reading vocabulary from ten words to fifteen represents a real achievement. The nature of this achievement, and of subsequent "growth spurts" is not lost on the semilog chart, which allows us to see clearly that the growth ratio between ten and fifteen words (5 divided by 10 = 50%) is equal to that of the difference for a slightly more advanced reader of adding 25 new words to an established vocabulary of 50 (25 divided by 50 = 50%). This is an interesting way to view learning.

Facilitating straight line analysis. A second characteristic of the semilog chart is that it tends to straighten out a line which, when plotted on an arithmetic chart, is curved—making it easier to analyze and interpret data. In order to see what we mean, contrast data concerning a child's performance in mathematics that have been plotted on both arithmetic and semilog charts (Figure 9:10). Suppose this child is just learning the skill, and begins by giving one correct response. Then he progresses to two correct, and from two to four. As he really catches on, he improves from four to eight, and so on. The general trend of this performance increase as it is shown on the arithmetic chart in Figure 9:10a, takes the form of an ever-steepening curve. That's because the data described here are increasing at a *constant percent*, or doubling each day. On an arithmetic chart, constant percent of increase (or any growth trend approaching it) appears as a curved line like the one seen on the arithmetic graph in the figure.

Look at the same data plotted on a semilog chart (Figure 9:10b). Here, the curve is flattened out to form a straight line, because the semilog chart shows data which increase at a constant percent as a flat line. This flattening of the curve makes it easier to describe the performance change currently taking place in terms of a trend line, and also to visualize what the future data trend might be if change continues in the same manner. Although this is oversimplifying a data analysis procedure explained later, we can lay a ruler along the line created by joining the data together and predict exactly where performance will go and how fast it will get there.

This magic act—turning a curve into a straight line—is possible because logarithms compress data. When data are plotted on a semilog chart, any curve that might exist on an arithmetic chart tends to be linearized, at least to a point where it is easier to analyze.

Flattening data bounce. The same flattening principle holds true for data that show a lot of "bounce." Figure 9:11 contains two charts, again arithmetic and semilog, on which identical data have been plotted. In this case, the child's performance, rather than showing a steady increase, makes frequent excursions, sometimes leaping up, then falling down, only to rise suddenly again (Figure 9:11a). Ordinarily it would be difficult to describe what is happening. Your eye is caught and held by each alternating peak and depression. On the semilog chart (Figure 9:11b), however, although data are still somewhat erratic, they are smoothed out enough to make an overall statement of direction and trend much easier to determine at first glance. Fluctuations are seen as much smaller than the arithmetic chart would seem to indicate, because they are represented in terms of percent of change.

FIGURE 9:10a. *Constant Percent of Change on Arithmetic Chart.*

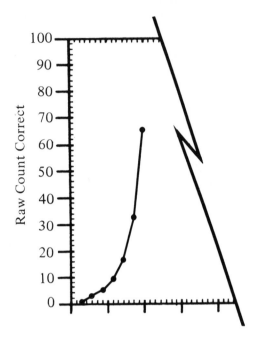

FIGURE 9.10b. *Constant Percent of Change on Semilog Chart.*

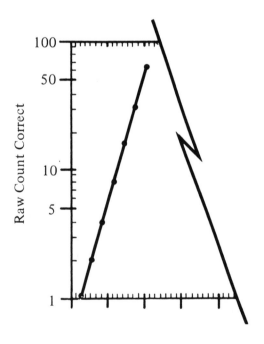

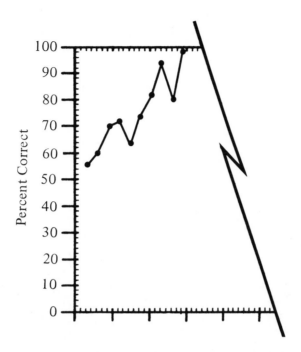

FIGURE 9:11b. *Differences in Representing Data Bounce: Same Data Bounce Plotted on Semilog Chart.*

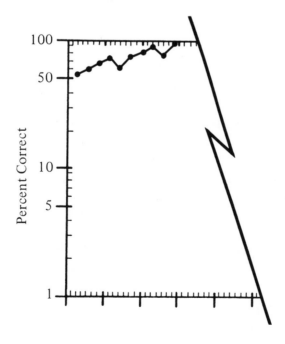

The Semilog Chart Format

This chart is called "semilog" because only one of its scales—the vertical—has a logarithmic basis. The horizontal scale is constructed much like that on any arithmetic chart.

The vertical amount scale. At first glance, the vertical amount scale is slightly intimidating, but it's really very simple to figure out. Scale intervals start large at the bottom, then get smaller as they go up. The effect is a little like an accordian being pulled open and then squeezed closed at one end. On the chart in Figure 9:12, this funny distribution of lines repeats itself three times, producing three sections of identical ruling.

FIGURE 9:12. *The Vertical Amount Scale on a Semilog Chart.*

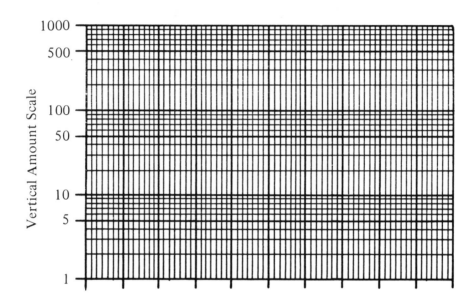

These sections are called "cycles" or "decks." Printed sheets of semilog paper may carry one, two, three, or even more cycles to suit the user's purpose. Let's look at the amount scale for a chart with just one cycle (Figure 9:13a), to observe in detail how a cycle is constructed. This scale starts with the number 1 at the bottom and proceeds upward through 10. You have already discovered that each line of the cycle is spaced progressively closer to the next. Find the 1 line; the 2 line; the 3 line; and so on, through the amount line marked 10.

As we have said before, equal distance on this scale shows equal percent of change. This particular cycle has been specifically marked to make one such proportional comparison very clear: The distances between the 1 and 2 amount lines and the 5 and 10 lines have been darkened to emphasize that the distances between these sets of points on the scale are equal, as are the proportional differences between 1 and 2, 5 and 10 (i.e., the increase between 1 and 2 responses is 100 percent, and the increase between 5 and 10 responses is the same). In addition to these comparisons, you'll find a number of other proportionate comparisons are easily made within this cycle. Try comparing the space between the 1 and 3 lines with that between the 2 and 6. Just as the ratios of 1:3 and 2:6 are equal, so are the distances between these amounts equal. Comparing the same distances on a similar portion of the amount scale for an arithmetic chart (Figure 9:13b) demonstrates again the marked difference in semilog and arithmetic chart construction.

FIGURE 9:13a. *Comparing Vertical Scales on Equal Interval and Semilog Charts: One Cycle Semilog, Scale 1 to 10.*

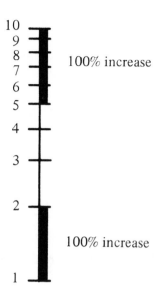

FIGURE 9:13b. *Comparing Vertical Scales on Equal Interval and Semilog Charts: Arithmetic Scale, through 10.*

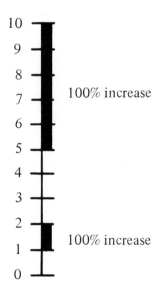

With a little practice, percent of change can be read directly from the chart. For instance, just as the proportionate difference between 1 and 2 can be calculated to be 100 percent, the same calculations can be applied to the amounts 2 and 3, arriving at a figure of 50 percent (that is, 3 is half again, or 50 percent more, than 2). This difference is represented on the chart by distance: the distance between 2 and 3 is 50 percent as large as the distance between 1 and 2. Figuring proportionate changes going up the scale (e.g., the difference between 3 and 4 is 33.3 percent increase; between 4 and 5 is 25 percent increase, and so on) results in the percentages marked on the amount scale in Figure 9:14.

Semilog charts may contain any number of cycles. Each additional cycle looks identical to the one preceding it, with the pattern of ruling repeated exactly. And, since equal distance means equal percent of change, the percentages found on Figure 9:14 may be applied to behavior increases or decreases no matter where on the chart they take place. Using the chart in Figure 9:15, which contains two cycles numbered from 1 to 100, you can easily test this for yourself. Mark with a pencil the distances between 1 and 2; 5 and 10; 10 and 20; 15 and 30; 20 and 40; 25 and 50; 40 and 80; 50 and 100. In each case, the distance marked off is the same. In a similar manner, the distance from 10 to 11 (10 percent increase) is equal to the distance from 20 to 22, or 50 to 55. This easy comparison is possible only as a result of the logarithmic basis for the amount scale.

By now you have realized a second feature of the semilog chart; that is, that *there is no zero anyplace on the vertical scale*. This is because of the logarithmic basis for the scale's construction. There is no logarithm of zero, and, therefore, zero cannot appear on the chart. The exclusion of zero on a behavior chart makes some philosophical sense. A count of zero (you didn't observe the behavior to happen at all) is always relative to the size of the sample you are taking. Suppose that you see "zero" talk-outs in a 20-minute sample. There is nothing to guarantee that the child won't produce a talk-out the second the sample is over. And this is true of any sample, theoretically, unless you follow a youngster around with a tally sheet for the rest of his life. Just because you don't see a behavior occur, does not mean that it doesn't occur at some time when you are not observing, hence, sampling, behavior. For this reason, "zero behavior" can never truly be said to occur, and zero really has no place on the chart. In all practicality, however, we must admit that there are plenty of times when you don't see a behavior occur within the time chosen for sampling and you want to say so on the chart: Simply pick a place beneath the lowest number you'll be using to record your data on the scale and designate it "zero." (For further detail on this issue, see the discussion on record floor in the next chapter.)

The horizontal scale. In contrast to the chart's vertical amount scale, the horizontal scale is equal interval in nature. This is because the scale represents time, which is considered to progress by addition, with each day added to all the time that has gone before it, rather than figured as a percentage of the past. The time accounted for on this scale may be in terms of sessions, days, weeks, or even months (although charting monthly would, in most cases, give us too little data to make adequate lines of progress—a procedure described in the next chapter). The horizontal scale on the chart in Figure 9:16 is identical to one which might be found on the arithmetic charts you are already familiar with. It is constructed to show successive calendar days, with each interval line representing one day. Darkened lines show the end of a counting cycle: in this case, a dark seventh day line, to signify the end of a week.

Entering Data on the Semilog Chart

As with the arithmetic chart, data can be stored on record sheets and entered on the chart whenever it is convenient. The more frequently you chart, of course, the more immediate is your access to what the child is doing. You can quickly verify those "feelings" about how an algebra program is working, or whether the out-of-seat contingency you have set on Clancy is really effective, each time you chart your data.

FIGURE 9:14. *Semilog Scale Marked to Show Percent of Change.*

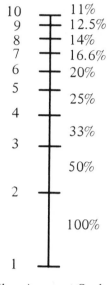

Semilog Amount Scale

FIGURE 9:15. *Two Cycle Semilog: Same Distance = Same Percent of Change.*

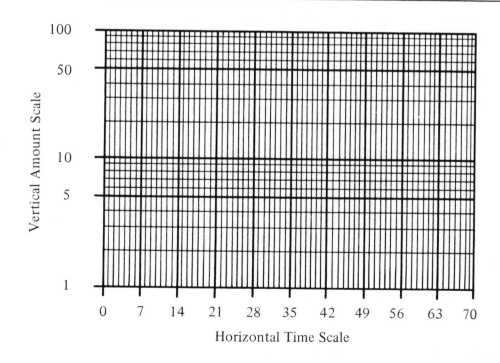

FIGURE 9:16. *The Horizontal Time Scale on a Semilog Chart.*

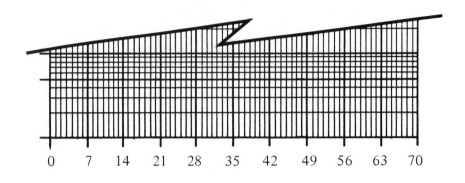

Successive Calendar Days

The steps used to enter data on the semilog chart are identical to those used for plotting data on an arithmetic graph. We'll give you some practice using the chart to display raw score data collected for two weeks on a youngster's worrisome *quarreling with peers* behaviors.

Step 1: Figure the data. Since these are raw score data, and to be charted as such, no calculations are necessary. The behavior counts (Figure 9:17a) may be entered "as is" on the chart.

Step 2: Locate the right amount interval. Your first job with a semilog chart is to find the right cycle in which data will be plotted. Since the scores on this record sheet range from between 2 and 30, you will need both the first and second cycles of the three-cycle chart shown in Figure 9:17b. The first week's data fall entirely within the first cycle, while those in the second week will be plotted in the second cycle.

You'll notice that only the heavy lines on the amount scale (1, 5, 10, 50, 100) are numbered. This is a kind of shorthand which has been adopted to make the chart's scale less messy, since some of the interval lines fall quite close together. Finding unnumbered lines may take a bit of initial practice, but it is not at all difficult. The 2 line, for instance, is located one line up from the interval numbered 1. Similarly, the 20 line is one up from 10. To locate a plot such as 25, find the 20 and 30 lines and make a mark slightly above half-way between them. A score of 28 would be slightly closer to the 30 interval line, and so on. Don't worry if you are not altogether precise in the placement of every score on this chart. You are interested in percent of change, and the percent of change involved in a plotting error such as locating a score of 32 at 33 is relatively slight. As long as your data plots are fairly close to the mark, and in relation to each other, the picture you get will be an accurate one.

Step 3: Locate the right time scale interval. This chart is calibrated according to seven-day weeks, with Monday, Wednesday, and Friday day lines signified during the first week, to make them easier to find. The first day's data are collected and entered on a Tuesday (Figure 9:17c).

Step 4: Make the data plot. Locate the point at which amount and day lines intersect and make the data plot (Figure 9:17d).

Step 5: Connect the plots. When you have charted all the correct data points, connect them in sequence with straight lines, going from left to right; first join points 1 and 2, then points 2 and 3, and so on, until all the data points are connected (Figure 17e).

FIGURE 9:17a. *Entering Data: Data Sheet and Raw Score Behavior Counts.*

Weeks	Mon	Tues	Wed	Thurs	Fri
1		2	5	4	8
2	12	20	25	22	30
3					

FIGURE 9:17b. *Entering Data: Locating the Amount Scale Interval.*

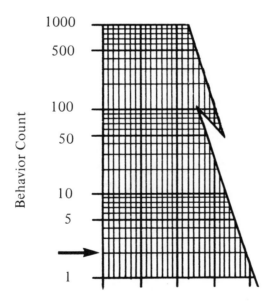

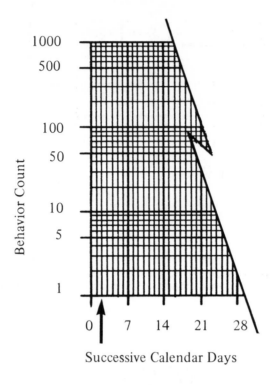

FIGURE 9.17d. *Entering Data: Making the Data Plot.*

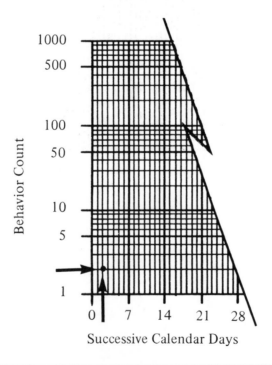

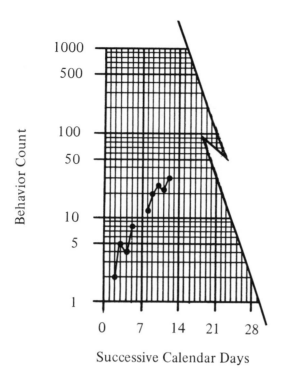

CONCLUSION

Whether you decide to choose arithmetic or semilog chart paper will depend upon a number of factors. In one sense, your choice can be based simply upon which chart you feel most comfortable using. In a deeper sense, however, your choice goes way beyond the aesthetic, practical, and comfort-giving aspects of these charts, for each chart presents a widely different picture of performance change. On the arithmetic chart, performance growth appears to be an additive and absolute entity. On the semilog chart, growth is expressed as proportional in nature. These fundamental differences are reflected in your selection of a chart, which must ultimately be based upon your own philosophy about the nature of learning.

Chapter 10
Making Charting Easier

A good chart is like an honest friend. It blows the whistle on the program that just isn't working; but, on the other hand, it is the first to congratulate you when you've hit instructional pay dirt. There's no way to avoid the fact, however, that charting is an activity which takes extra time and effort. Perhaps you relinquish that time just a little grudgingly. You could, after all, be making up next month's lesson plans, setting up materials for a really spectacular art project tomorrow, reviewing a new set of curriculum materials, or even just catching up on last week's movie tabloid. Nonetheless, we feel that charting is well worth whatever energy you can devote to it.

But, while a little work is inevitable, you shouldn't have to do more than necessary. The following are a few simple hints aimed at making charting less arduous, and the process of reading and using data (what the next chapter's all about) much easier.

GETTING READY TO START CHARTING

Using Chart Conventions

Charting—like sports, politics, and good conversation—observes certain rules of etiquette which keep things functioning more smoothly and ensure that all the participants communicate effectively. Experienced charters call these rules "conventions." Conventions are developed to standardize and lend uniformity to the charting experience. But let us stress that the conventions listed and described in this section, while popular, are not intended to dictate the way you set up your charts. These are only ideas collected from journals, workshops, and practical settings in psychology and education, where individuals have used charts for many years to describe how children behave and change. There are other conventions, and other ways of indicating the same information which have not been included. Read through the information presented here critically, to decide which conventions fit your charting style and needs and which do not. You may choose to use a few, many, or none of these conventions, but try, whatever your choice, to keep your own charts uniform. The effort will make data interpretation and comparison easier for all.

Conventions for labels and legends. The following conventions, developed by the American Psychological Association (Katzenberg, 1975), will enable you to keep certain "set-up" information on the charts you design for yourself consistent and easy to read. They allow all who see the chart to understand basic information immediately, such as what behavior and time variables are being charted, what interval units are employed, and how behavior components (e.g., correct and incorrect) may be identified.

— *Labeling axes.* Label axes clearly with both the variables being measured and the units in which they are measured. *Print* labels, in preference to cursive script. In general, abbreviations are not punctuated (e.g., MIN, SEC, etc.). Label the independent variable (DAYS, SESSIONS, WEEKS) on the horizontal *x*-axis, and the dependent variable (the child's responses) on the vertical *y*-axis (Figure 10:1). Printing should be clear, parallel to the proper axis, and centered.

FIGURE 10:1. *Labeling the Chart Axes.*

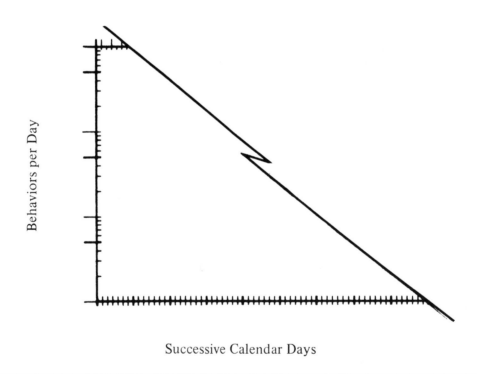

Successive Calendar Days

— *Entering scale values.* Assign numerical values to each of the grid points drawn. Values should represent convenient scales. Numbers must be clearly written and easy to read (Figure 10:2).

FIGURE 10.2. *Entering the Scale Values.*

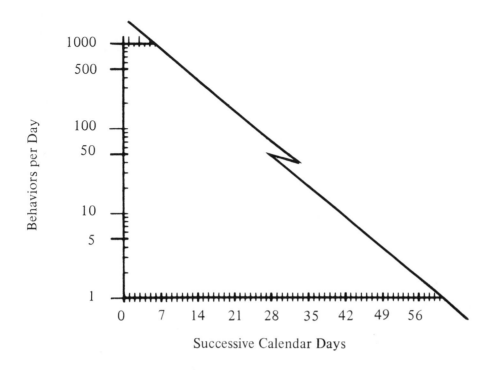

Conventions for entering data on the chart. As we stressed at the outset, using a set of standard symbols for the various pieces of information entered on the chart makes the charting process faster and interpretation easier. Keeping symbols consistent from chart to chart can make the difference between an undecipherable constellation of dots and lines, and a clear, concise, precise record of the child's growth. The more clearly you are able to see the child's progress, the more timely and appropriate are the instructional decisions you are likely to make.

The following set of symbols and charting conventions has been developed and used over a number of years by teachers for charting the social and academic performance of children in their classrooms. Many of the concepts and symbols (e.g., record floor, record ceiling, ignored-, data-, and no chance days) originated in Precision Teaching classrooms. While they have proved invaluable in helping teachers organize, analyze, and make programming decisions regarding countless pieces of behavior data, you may not find them applicable to your own. *Record floor*, for instance, is, as you will see, a concept applying to only certain types of data. Whether or not you decide to use specific conventions, the concepts underlying each of these symbols will give you a greater understanding of the variables which influence our picture of behavior change.

— *Data plots.* When more than one set of data appears (e.g., correct and incorrect data on the same chart), some means must be provided for distinguishing them from one another. If you don't care about reproducing the chart directly from the original, you can use a different color for each set of data and add a key, indicating what each color represents. Color lends considerable attention value; but, if you intend to xerox the chart, remember that colors will lose their effectiveness on the copy. A better way to distinguish between variables is to use different symbols for each data set; for example, dots for corrects and triangles for incorrects. The line connecting the data points can also vary, using solid lines, dashes, dots, or combinations of dots and dashes. Figure 10:3 shows a portion of a chart on which correct data have been plotted using dots, and error data with small *x*'s.

FIGURE 10:3. *Representing Data Plots.*

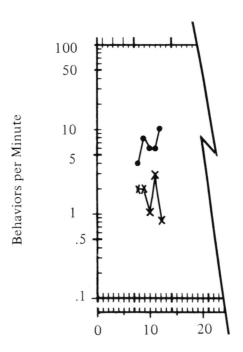

— *Data days.* Any day on which data are taken and the results recorded on the chart is a *data day*. All data day plots are connected consecutively by a line drawn from left to right, unless a *no-chance* or *ignored day* intervenes.

— *No-chance days.* These are the days on which there was no chance for the behavior being counted to occur in the situation in which data are usually taken. Weekends, vacations, and school absences are no-chance days, as are days on which a special activity, such as a field trip, prevents data collection. To emphasize no-chance days, not only are no data charted, but the day lines are left totally blank; that is, data points are *not* connected with a line across them (Figure 10:4). Leaving these days blank makes it easy to see the effects of absences or no practice. Sometimes cycles are obvious: Weekends might always be followed by a dip in performance rate which is overcome by Friday, but repeated again on Monday. The discovery of patterns like these should prompt you to examine procedures which might alleviate such performance cycles (e.g., initiating extra practice at home to help the child progress more evenly).

— *Ignored days.* Sometimes there are circumstances under which behavior occurs, but for one reason or another the count is not recorded. Maybe you lost the tally sheet, or perhaps you simply don't have time to count *out-of-seats* on a particular day. These days are treated differently on the chart from a no-chance day, to allow you to discriminate that the chance for behavior did occur. Although no data are plotted for the day, the two data days on either side of it are connected by a line drawn across the ignored day (Figure 10:5). It's important to be able to discriminate between ignored and no-chance days. While a child might show a pattern of data decline after weekends and other days on which he has no chance to practice or perform, he may show an improvement after an ignored day—you didn't catch the data, but the child nevertheless had an opportunity to behave. If you repeatedly find that improvement occurs even after ignored days, you may decide that collecting data isn't necessary each and every day, as long as the opportunity for performance to occur is there.

— *Record ceiling.* The record ceiling represents limits which our own instructional procedures impose upon a child's performance capabilities. Such a ceiling may be established in many ways. Some measures carry a built in ceiling—percent, for example, which always imposes a ceiling of 100 percent. A ceiling may also be a function of the way in which a learning task is structured.

To gain a first-hand understanding of the notion of record ceiling, try this with a group of friends. Give each person a pencil and a blank sheet of paper. Then tell them that you will give them one minute to make as many slash marks on the paper as they can. When the minute's up, check the papers. If your friends are like most groups we have asked to do this simple task, you will be surprised at the range of behavior you get —all the way from a small number of slashes to sheets covered with small marks reaching into the hundreds. Imagine the same group now as you tell them to make marks only as you direct them. With a cue to make a slash only every two seconds, your friends who had highly individual performance patterns on the first go-around now end up looking very much alike; all papers have 30 slashes. The ceiling you have established is 30 and no one's performance can exceed this number.

The same principle is easily observed in the way many classroom programs are set up (e.g., you never present a worksheet with more than ten problems on it; or you never give more than 20 flashcards to introduce new vocabulary to a child). In certain instances, ceilings may be necessary and most desirable. It's important to recognize, however, where such ceilings do exist, and to understand the limitations they will place on the range of performance you can expect to see. The performance expectations these ceilings automatically impose may be much more or much less than the child is actually capable of doing.

Record ceiling is designated on the chart by a dashed line drawn across the appropriate day line (Figure 10:6). For instance, the semilog chart in Figure 10:6 has record ceilings entered at 15, 30, and 25 to show that these were the respective numbers of problems assigned on each of the three days performance data were collected. The ceiling *floats*, or changes from day to day. On some charts, the ceiling is actually the upper limit of the chart's vertical amount scale, and no ceiling indicator is needed. This is true in the case of percent where 100 percent at the top of the scale marks the upper limit of correct performance, as well as on a raw score chart designed to correspond exactly to the number of problems assigned each day.

For analysis purposes, it's helpful to keep a ceiling which is constant from day to day. Figure 10:7 shows a *stable* record ceiling at 30 problems assigned each day. The data plots show clearly that performance is growing steadily closer to the maximum (i.e., 30) established by the teacher.

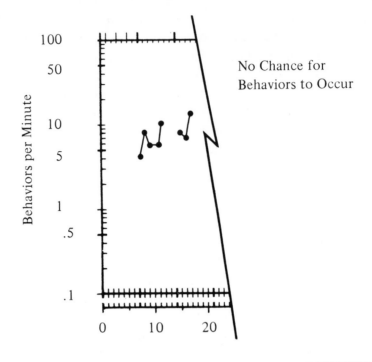

FIGURE 10:5. *Ignored Days.*

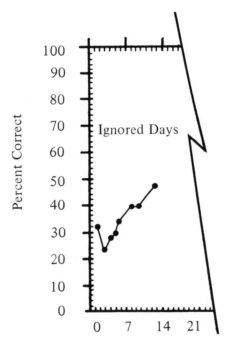

FIGURE 10:6. *Floating Record Ceiling.*

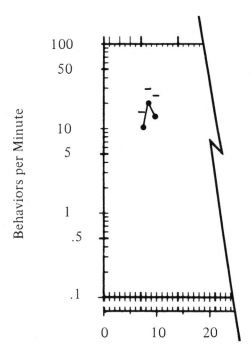

FIGURE 10:7. *Stable Record Ceiling.*

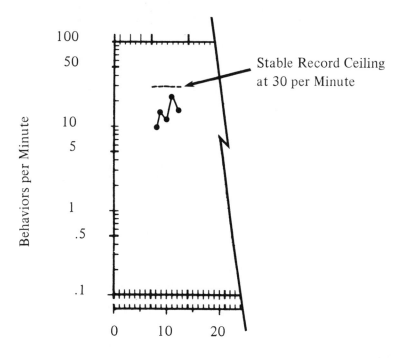

— *Record floor.* A concept developed primarily for use on the semilog chart, record floor designates the value of the lowest measurable or recordable amount; that is, where one occurrence of a behavior would appear on the chart.

How this value is determined varies from measure to measure, but how it is recorded on the semilog behavior chart remains the same, no matter what measure is used. The record floor is shown by drawing a horizontal line through the vertical day line for each day on which data are gathered and entered. Figure 10:8 shows the record floor entered for data collected in a one minute sample each day. The horizontal line is drawn through successive day lines at the place where *one behavior per minute* would be recorded. When the record floor remains constant from one day to the next, a single line can be drawn at the appropriate point, connecting across all the data days in question for each successive period (e.g., a week) over which data are collected. Record floor is an especially important concept for all plotting done on the semilog chart, and for the arithmetic as well, in those instances in which percent is the measure charted.

FIGURE 10:8. *Record Floor.*

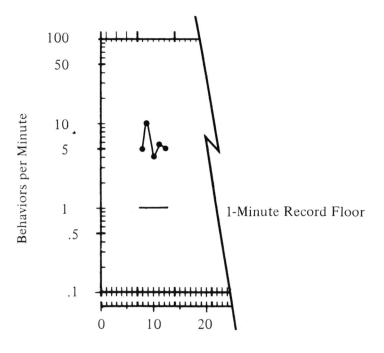

The record floor concept originated in part due to a need to designate where "zero" should lie when rate data were charted on semilog paper. Remember, the semilog chart, unlike the arithmetic graph, has no zero designation on the amount scale. The only way to record a behavior count of zero on semilog paper, then, is in reference to the record floor. Since the floor indicates where one occurrence of behavior would be charted, anything plotted below the floor is read as "zero." Pick an arbitrary point just below the record floor line and make a data plot. In Figure 10:9, the x-mark below the record floor shows that zero performance errors were made on the data day in question.

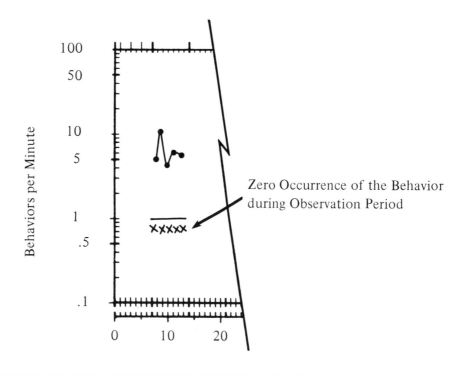

Entering a record floor requires first, of course, that you calculate it. How the record floor is calculated for a semilog chart depends upon the measure you are charting. When raw score is the measure, the record floor is always *1*, and designated at the "1" line on the chart. If rate data are being recorded on a semilog chart, the floor value is calculated by dividing the lowest non-zero score (i.e., 1) by the total amount of time observation takes place. Determining the record floor on a one-minute sample is easy: 1 divided by 1. For data taken over a ten-minute session, on the other hand, the floor is at .1 (1 divided by 10). Data taken over a 100-minute session would dictate a floor at .01 (1 divided by 100). A two-minute sample would result in a floor of .5 (1 divided by 2); in a 20-minute sample, the floor is .05. Record floors for one-minute, two-minute, three-minute, four-minute, and five-minute samples are shown on the semilog chart in Figure 10:10.

Although designed initially for the semilog chart, the record floor also serves a special function on an arithmetic percent chart. Here, it's an immediate visual cue to how many problems or items over which performance is being measured. Such information is often useful when performance characteristics such as endurance are an issue, and helps in interpreting data changes seen on the chart over time. The floor will vary considerably, for instance, depending upon whether two, three, four, or more problems are assigned. To figure the record floor when only two problems are given, divide the lowest possible amount of measurable correct performance (one problem) by the total number of problems possible (in this case, two problems). The resulting percentage (50 percent) represents the value of the lowest possible measurable performance score. The child could achieve only two possible correct scores under these circumstances, 50 and 100 percent. When three problems are assigned, the record floor is at 33.3 percent on the chart (one divided by three). For four items, a record floor of 25 percent (one divided by four) is obtained. As more and more items are assigned, one response becomes a smaller percentage of the whole, and so, of course, the record floor is lower. The nearer a record floor approaches zero on an arithmetic percent chart, the more items over which performance is being measured.

FIGURE 10:10. *Sample Record Floors on Semilog.*

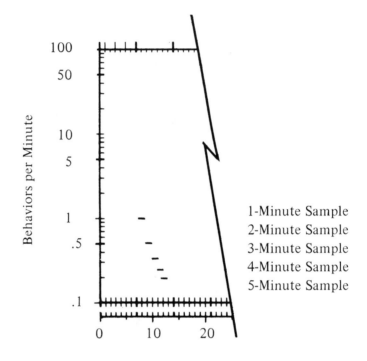

If no record floor is marked on the chart, data may sometimes be tricky to interpret. Say that a percent chart shows three days' data: 25 percent for the first two days, and 50 percent on the third. Correct performance seems to be improving, yet it is important to note that on each of the first two days four problems were given (25 percent record floor) and on the third, two problems were assigned. The child got only one problem correct on all three days. Were the record floor entered (Figure 10:11), you would be able to see that a 25 percent score is not possible on day three, and thus make some very different assumptions about performance.

In cases such as this one, the record floor comes in handy by reducing the chances that we will make erroneous comparisons about data taken under widely diverse conditions. In rate, for example, the amount of time allowed for performance is an important endurance issue. A child could hardly be expected to perform at the same rate over a 20-minute period as he did in a one-minute sample. Noting the record floor on the chart, it is possible to discriminate which were short samples, and which longer; thus, we may arrive at one possible explanation for dramatic data fluctuations.

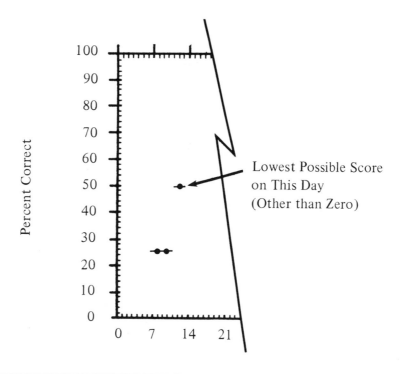

Whenever record floors drift considerably from day to day, accompanied by data entries that fluctuate to follow the bouncing floor, data interpretation becomes more difficult. Keeping the floor consistent over several days removes one extra confounding variable from the decision making process. In Figure 10:12, both correct and incorrect data are recorded in conjunction with a stable one-minute record floor. When a performance ceiling is added (Figure 10:13), the outside performance limits established by the teacher are clearly visible: the child is given one minute each day to do as many problems as he can of a total 20 problems assigned. The chart in this figure shows that the child's correct performance travels, for the most part, steadily upward, while errors are fairly consistent at zero.

FIGURE 10:12. *Entering Data with a Stable Record Floor.*

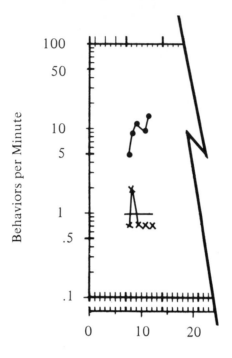

FIGURE 10.13. *Record Ceiling and Record Floor.*

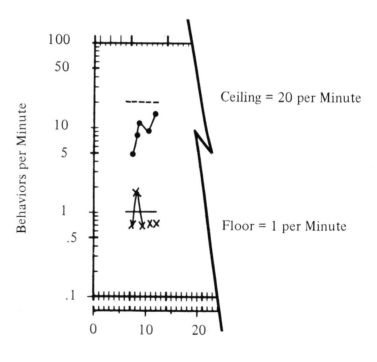

— *Phase change lines.* These lines are drawn on the chart to signify each time a change is made in the teaching program. Phase change lines are drawn vertically, in the space just preceding the day on which the change first takes place (Figure 10:14). Data plots are *not* connected across phase lines, in order to make clear what patterns, if any, are associated with the plan changes.

Notes explaining the nature of each phase change are written above the data, horizontally along the chart, under a phase "umbrella" which extends from the phase change line itself to the end of the phase. Notes may refer to changes in the instructional plan, the introduction of new material, or anything that will explain briefly and clearly what kind of event changes have been initiated which might be responsible for future changes in the child's performance.

FIGURE 10:14. *A Program Phase Change.*

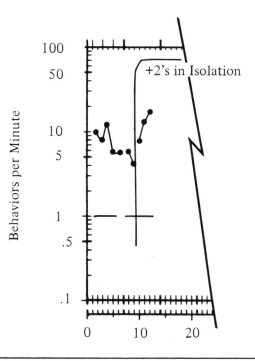

— *Aim.* The objectives we define for programs are written with a varying degree of specificity. In Chapter 4 a number of components for objectives were outlined, which included behavior, content, condition, criterion level and expected completion date. While an objective need not contain all of these components, an aim is a very specific kind of objective. The aim must always contain a criterion level and for best results should also include an expected completion date.

The aim symbol is one of the most important, since it is a graphic representation of the direction you hope your program will take. The aim is entered on the chart with a symbol that looks something like a star. Form the symbol by drawing an arrowhead with a line through it at the point where the aim amount and aim date coincide (Figure 10:15). If the aim is an acceleration target, the arrow points up (🔺); if the aim is a deceleration target, the arrow points down (🔻). The aim acts as a reminder to you, from day to day, where your data are supposed to be heading—a real help, since it's easy to get swallowed up in the daily routine and forget exactly where it was you wanted the child to go.

Suppose you set an aim for multiplication facts at 30 per minute. Place the aim mark on the line representing the intersection of the desired amount of behavior (30 movements per minute) and the aim-date. Aims can be specified for both correct and incorrect performances.

Aims are not inflexible. Sometimes teachers set unreasonable aims, and sometimes children fail to meet our most carefully formed expectations. If it becomes apparent that it is not possible for the child to meet the established aim, within the expected time, reset the aim at a new and later date.

FIGURE 10:15. *Representing Performance Aims.*

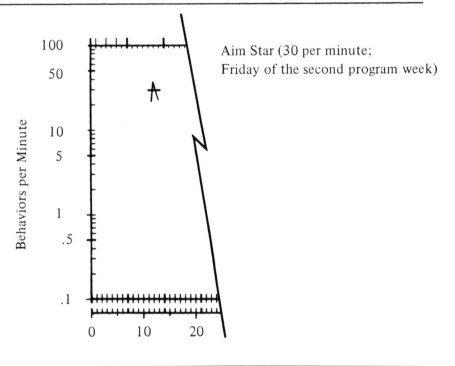

Aim Star (30 per minute; Friday of the second program week)

Figure 10:16 shows two charts, one semilog and one arithmetic, on which two separate sets of data and accompanying information have been entered using the charting conventions just presented. (Note that record floor is not shown on the arithmetic chart, as raw score data are being plotted and the record floor is obviously located at "1.")

FIGURE 10:16a. *Semilog Chart with Conventions.*

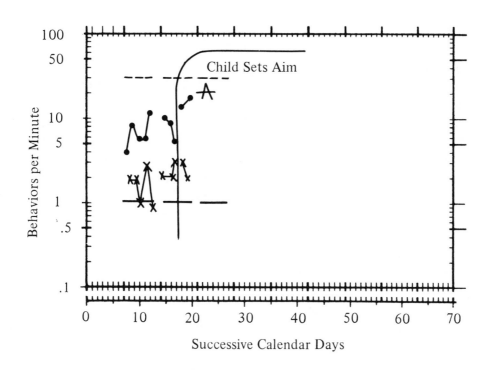

FIGURE 10:16b. *Arithmetic Chart with Conventions.*

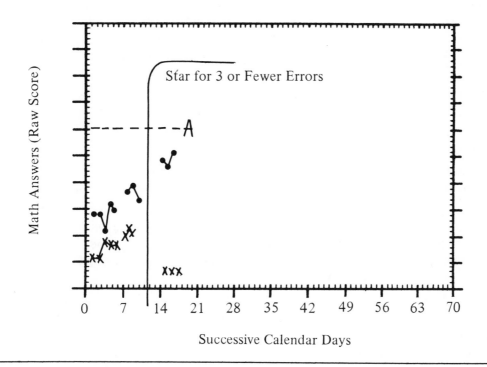

Using a Standard Chart

The more consistent and standardized the format used for measuring and charting, the more easily we can make and interpret charts. Charts that look the same are less confusing to enter data on, and less confusing to compare. You yourself can come up with a standardized chart form, no matter what kind of chart you prefer or what measure you choose to record.

A standard arithmetic chart. An example of a standardized percent chart is contained in Figure 10:17. You might wish to make a stencil of the chart shown here and run copies for yourself and friends who chart.

Standardizing means, of course, making all features of the chart uniform for all the data which will be plotted. The amount scale is one of the most important of those features, and often one of the most difficult to standardize. When percent is the measure, the range on the amount scale remains constant—from 0 percent to 100 percent. But suppose you want a chart that can be used for any one of several measures (e.g., percent, rate, and raw score). Or, perhaps you need a chart which can accommodate data (e.g., raw scores) of children performing at several different levels (e.g., 0-10, 50-75, 100-200).

To make one chart fit all, you must create an amount scale with a range broad enough to encompass performance at all levels. (This may cause certain interpretation problems later: For instance, a chart with a very large scale, say 1 to 300, tends to compress the appearance of data plotted at the lower limits of the scale. But such problems will be dealt with later.)

Other features which may be standardized include the time scale and notation of information concerning the pupil, behavior, program, and instructor or program facilitator. These will be covered in more detail in the section that follows.

A standard semilog chart. A standard semilog chart, developed specifically for child performance rate data, has been available commercially for quite some time. This chart, called the Standard Behavior Chart, was created several years ago by O. R. Lindsley. Based upon the accumulation and study of thousands of pieces of data on the academic and social performance of children, the chart has undergone rigorous field testing and refinement. The Standard Behavior Chart (one similar to it appears in Figure 10:18) has been used by teachers in regular and special education classrooms with children of all age levels across the country for over ten years. It's been a special favorite of teachers involved in what has been called "Precision" and, most recently, "Exceptional" teaching.

One of the reasons for the chart's popularity is its six-cycle amount scale, which offers a total charting range of .000695 to 1000. This range has particular advantages for the charting of rate per minute). Such incredible range and flexibility makes it possible to track, display, and compare every 24 hours) as well as those occurring at the very upper limits of human capabilities (e.g., 1000 per minute). This incredible range and flexibility makes it possible to track, display, and compare data on human behavior at virtually all levels of performance using identical charts. For instance, a teacher might want to look at two very different behaviors for a particular child. He might chart the youngster's math performance, timed during one-minute samples and occurring within a range of 1 to 65 responses per minute. At the same time, he might be interested in a social behavior problem such as swearing, which occurs frequently enough to be troublesome and yet, tracked over an entire 300 minute day, might fall in a very different range, say between .01 (once every 100 minutes) and .2 (2 every 10 minutes) when the rates were figured. Normally, the teacher would design individual charts with different scales for each of these behaviors. But with six cycles at his disposal, he can easily fit both behavioral pictures on a consistent chart form.

Comparisons of data for one child, or across several children functioning at different parts of the chart are very easy with six cycles available. In the case just cited, the teacher can compare what happens in math performance as swearing takes an increasing or decreasing trend, without

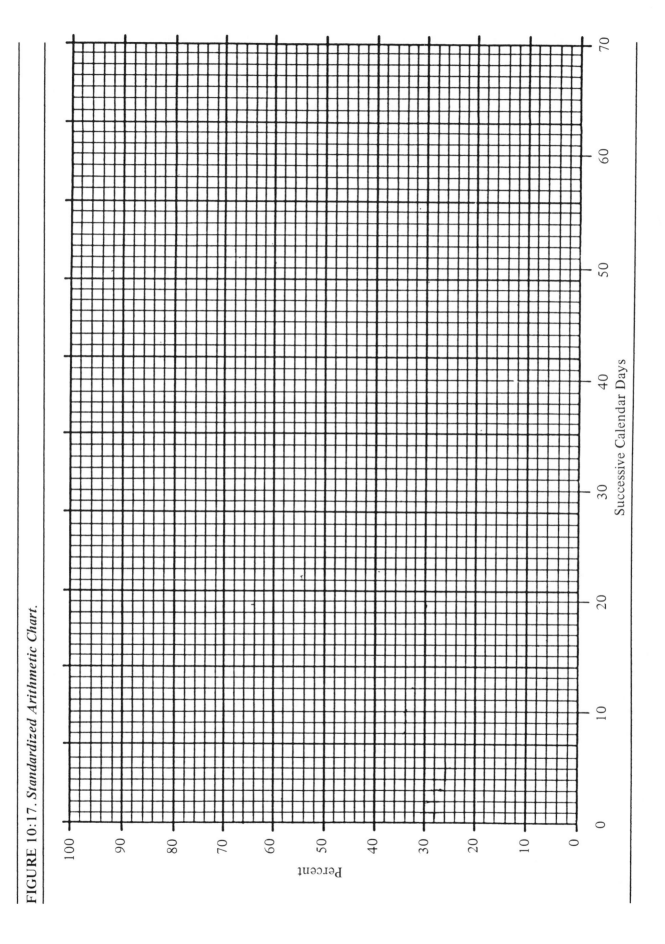

FIGURE 10:17. *Standardized Arithmetic Chart.*

FIGURE 10:18. *Standardized Semilog Chart.*

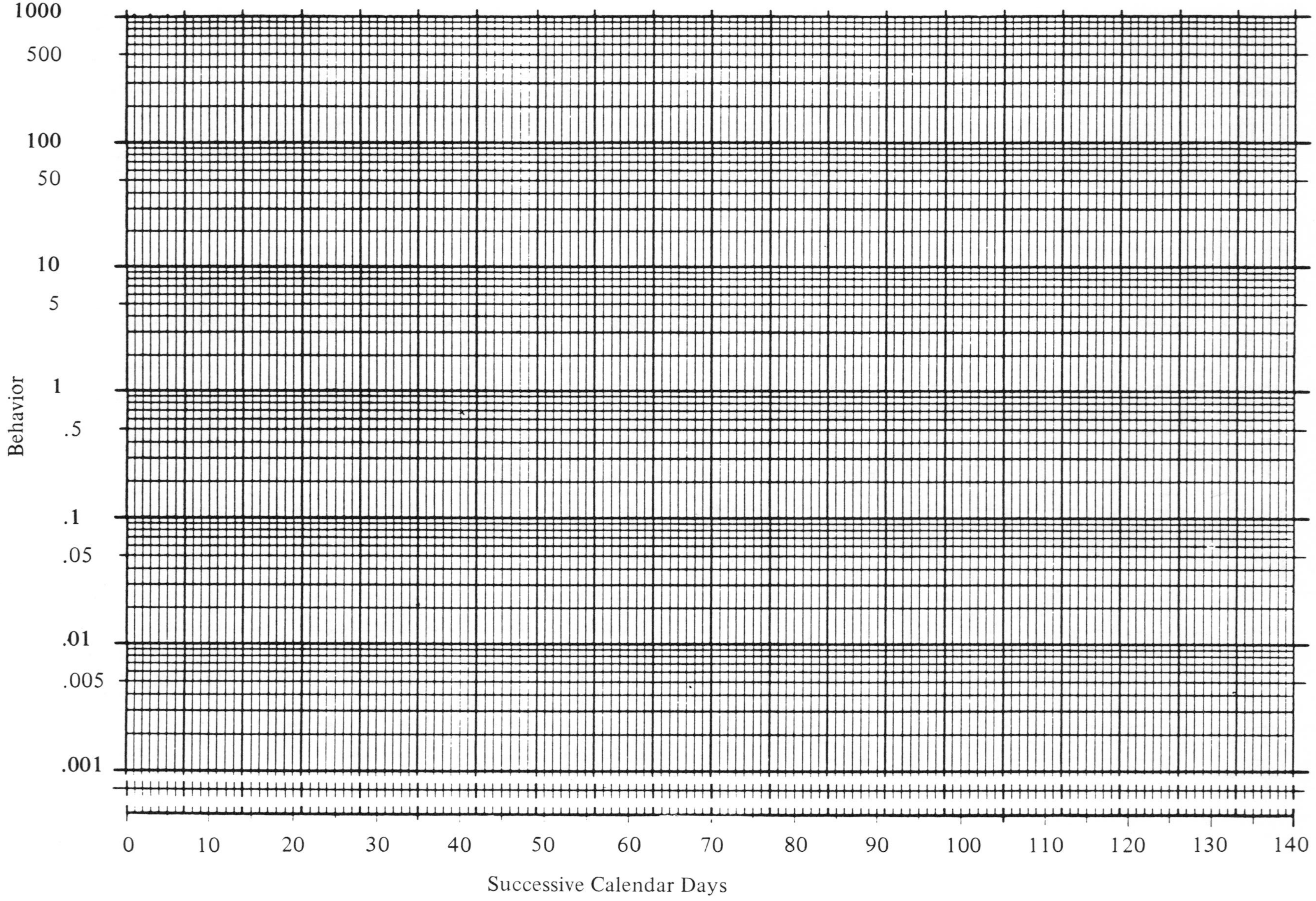

Behavior

1000
500

100
50

10
5

1
.5

.1
.05

.01
.005

.001

0   10   20   30   40   50   60   70   80   90   100   110   120   130   140

Successive Calendar Days

having to translate data from one chart form to another. It's much easier to make such comparisons using an identical chart format for all data, rather than readjusting thinking to suit a variety of different scales.

One further important advantage of having so many cycles on one chart is that it keeps teachers continually aware of the *total range of possibilities* for growth. Suppose, for instance, that you teach a classroom of moderately handicapped youngsters. You find a one-cycle chart quite adequate for tracking the addition fact performance of a certain child who usually does between one and ten facts a minute correctly. Eventually you become so accustomed to that single range of performance in front of you, that you forget there is any other. You may hold the child to less than he is capable of doing simply because you lose sight of the range of performance possible. Being constantly confronted by a wide performance range forces you to think about other dimensions of behavior and helps you to maintain a broader perspective on the performance you are viewing.

In order to keep the vertical scale from looking cluttered, only certain lines on the chart are numbered. Looking at the chart in Figure 10:18, you can see that all the 1 and 5 lines (i.e., lines representing a figure that includes the digits 1 or 5, such as .01, .1, 1, 10, 100, 1000, or .005, .05, .5, 5, and so on) are clearly numbered. The 1 lines are darkened to make the beginning of each cycle readily apparent. Because a semilog chart is constructed in cycles you are already familiar with, and each cycle's numbers are ten times the value of the numbers on the cycle immediately below, figuring out the values of the unnumbered data lines is easily mastered.

In addition to allowing charting of behaviors falling within an extensive performance range, the standard semilog chart, like the standard arithmetic graph, makes it possible to track those behaviors over an extensive period of time. The horizontal time scale contains 140 days, enough for an entire semester of data keeping. (Two charts will take you through a school year.) Sunday lines are darkened to help mark the end of each week. Monday, Wednesday, and Friday can be labeled on the appropriate lines at the top of the chart for the first week, to assist you in finding the correct day lines. In short, the chart is further standardized by being set up to allow calendar "locking," or coordination, so that all charts kept for a particular time period can be easily monitored and compared, with no confusion about when particular programs or program changes start or end.

Additional information important to later data transfer can be added in spaces which you may wish to provide at the bottom of the chart. Such information might include:

Child's Name............Teacher........... Supervisor........... Observer...........

Child's Age .... Classroom...............Behavior Observed...............

Description of Behavior..

....

All these features go together to make a chart that encourages consistent data keeping for a single teacher, as well as promoting data sharing among teachers and other individuals working with the child. Further, recording this information on the chart itself ensures that these details will not be lost as the conveniently sized 8½ x 11 chart is stored and filed for future reference.

A Charting Shortcut for Rate Data: The Rate Finder

If you are charting rate data, determining record floors, record ceilings, and calculating correct and error rates may be a part of daily data keeping you find unavoidable, but bothersome, nonetheless. All that division before you can lay a pencil to the chart! Happily, each of these operations can be performed more quickly and easily with the aid of a device called the Rate Finder, which will do all the computation for you.

A Rate Finder made for use with six-cycle semilog paper is printed in Figure 10:19 in full size. You can make a photo transparency of this Finder for use on your own charts. Notice that the

---

FIGURE 10:19. *The Rate Finder.*

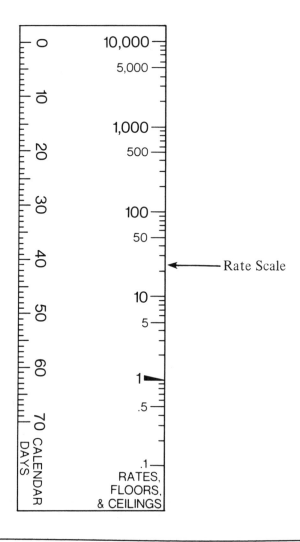

right side of the Finder has a logarithmic scale very similar to the one printed on your six-cycle chart. The Finder does its calculations by acting as a sort of slide rule in combination with the chart's vertical amount scale, to help you compute rate figures, record floors, and ceilings. All the division is done simply by moving the Finder up and down appropriately on the chart.

A day scale which is a virtual reproduction of the horizontal calendar scale found along the bottom of the Standard Semilog Chart printed in this book is also included on the Finder. This scale performs no calculations but it will allow you to tell at a glance how long a particular program phase has been in effect, as well as other time related information, such as the length of a vacation or absence, or how many days remain before the child needs to meet his aim.

Calculating rate record floors on a semilog chart. The record floor is generally the first entry on the chart (Figure 10:20). It serves as a sort of base to which all other data information—correct responses, errors, and ceiling—are anchored. To find the record floor:

— Place the Finder on the chart with the rate scale to the left of the day for which you wish to find the floor.

— Locate the number on the Finder's scale which is equal to the number of minutes that were spent counting behavior.

— Move the Finder up or down the chart until the number on its scale which is equal to the number of minutes is directly over the 1 line of the chart.

— Make a dash line next to 1 on the Finder's scale, crossing the appropriate day line. This is your record floor. If the record floors are to be consistent from one day to the next, remember you can fill them in now on the chart, drawing a straight line from the Monday to the Friday day lines. (It is to your advantage to have consistent record floors not only for the sake of interpretation as we pointed out earlier, but also for ease in entering data, as you will soon discover.)

Finding the record ceiling. If you desire to locate a record ceiling (Figure 10:21):

— Keep the Finder in the same position. Find the number on the rate scale that represents the total number of behaviors possible under the given situation. Make a dash line on the chart next to this point, indicating the record ceiling.

Calculating and locating rate. Rate data entries are the last made on the chart, using the Rate Finder (Figure 10:22).

— Still keeping the Finder in the same position, locate the number on the Finder's scale that represents the number of correct behaviors counted. Make a dot on the chart next to this number, indicating the correct rate. If no correct responses were counted, make a dot at an arbitrary point just below the record floor line. (Remember, there are no zeros on this chart.)

— Still keeping the Finder in the same position, locate the number on the rate scale that represents the number of error behaviors you counted. Make an x on the chart at that point. If no errors were counted, place the x just below the record floor line.

Other Charting Materials

Recording implements. The only implements necessary in ordinary charting are a No. 2 pencil and a small ruler or a straight-edge for connecting data points. A plastic template such as that in Figure 10:23 will act as a straight-edge, and at the same time make entry of charting symbols like

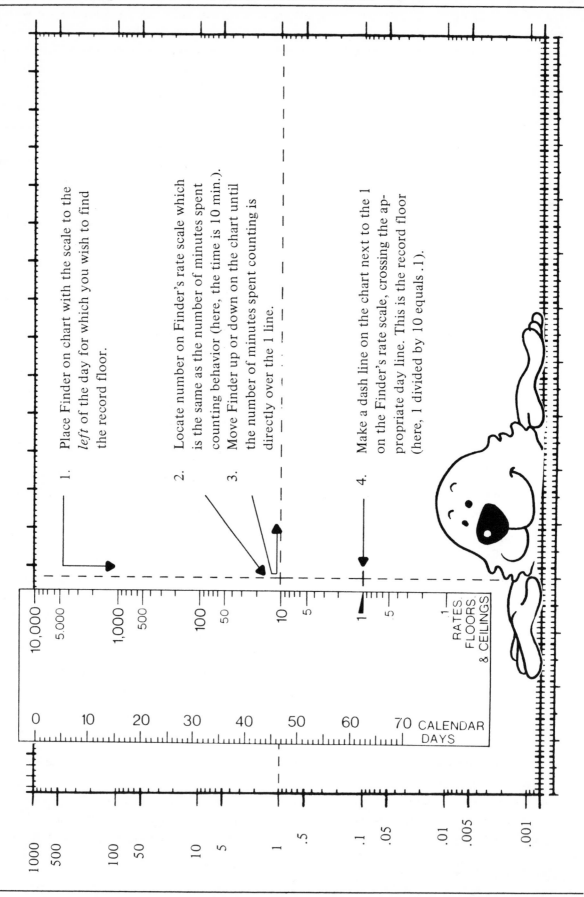

1. Place Finder on chart with the scale to the *left* of the day for which you wish to find the record floor.

2. Locate number on Finder's rate scale which is the same as the number of minutes spent counting behavior (here, the time is 10 min.).

3. Move Finder up or down on the chart until the number of minutes spent counting is directly over the 1 line.

4. Make a dash line on the chart next to the 1 on the Finder's rate scale, crossing the appropriate day line. This is the record floor (here, 1 divided by 10 equals .1).

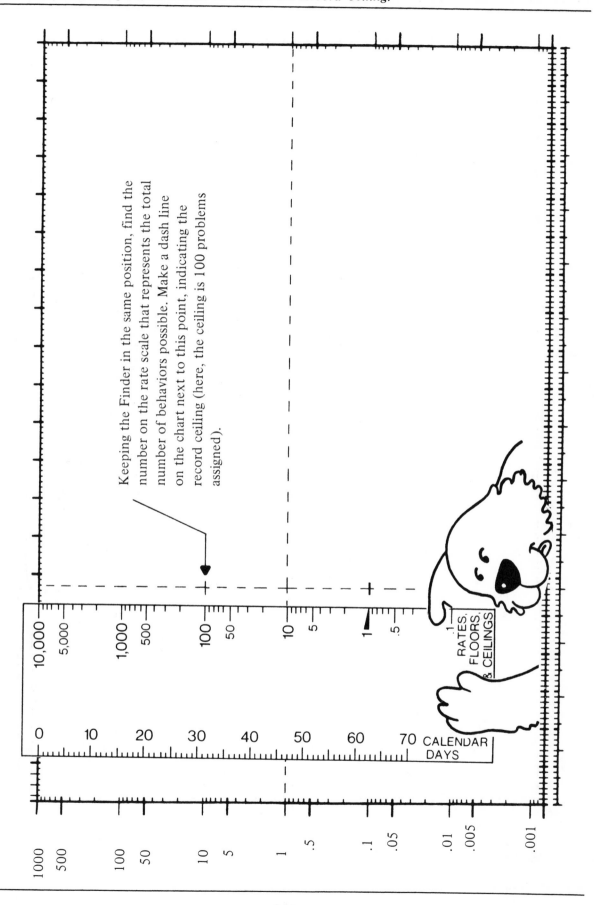

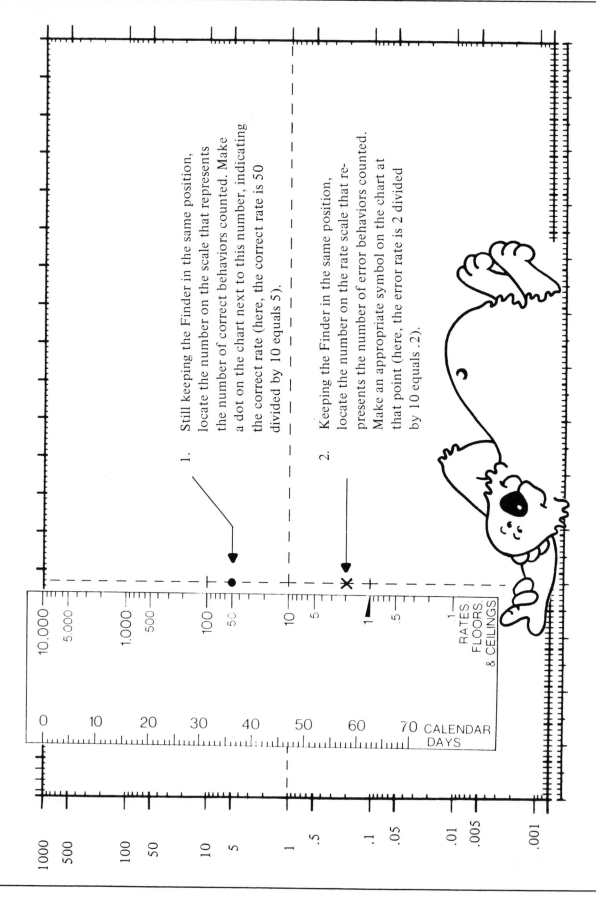

1. Still keeping the Finder in the same position, locate the number on the scale that represents the number of correct behaviors counted. Make a dot on the chart next to this number, indicating the correct rate (here, the correct rate is 50 divided by 10 equals 5).

2. Keeping the Finder in the same position, locate the number on the rate scale that represents the number of error behaviors counted. Make an appropriate symbol on the chart at that point (here, the error rate is 2 divided by 10 equals .2).

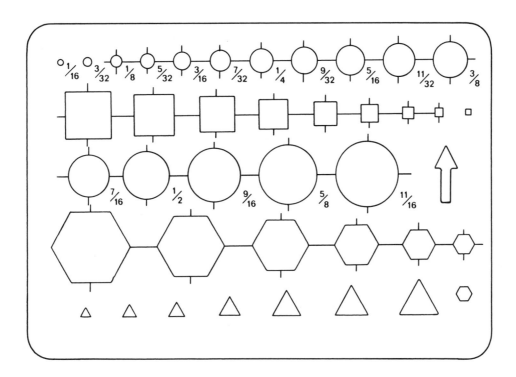

those already explained, and some to come in the next chapter, quicker and neater. The line showing the upward and downward movement of behavior on the chart should be heavy enough to stand out boldly on the chart, but neat as well. Original charting should never be done in ink because of the possibility of plotting errors.

Chart paper. If you choose not to use a Standard Behavior Chart, you will have to obtain some other kind of paper suited to your needs. You can make your own chart paper, but the operation is tedious and sometimes less than satisfactory. Ready-ruled chart paper is usually easily obtainable and generally not too expensive, at any stationery store or dealer specializing in engineering and business supplies.

As you select chart paper, keep these factors in mind:

— *Weight.* In general, light weight paper is best as long as the chart won't be handled too much, and especially if you expect to superimpose charts over one another for the purpose of comparing or copying the lines of progress. Heavier paper should be chosen if the chart will be handled quite a bit and durability is the main consideration.

— *Paper size.* What size chart you choose is most often a matter of personal preference and convenience. Sheets small enough to be fitted into a pocket notebook are available, but these are not really practical. Most chart paper is available in standard letter size, 8½ x 11, making it easier to file or keep in a standard three-ring binder. It is a good idea to have all charts uniform in size, for easy reference and in order that all can be stored together.

— *Chart color.* The color of the charting grid, too, is widely variable, and often left to the charter. A light color, such as olive green or light blue, is easier on the eyes, especially under artificial light, than other colors. A line of progress done in almost any color of pencil will stand out prominently against grids of such color.

— *Chart reproduction.* Often you will want copies of a particular chart to save for future use, to send to the child's next teacher or to other interested individuals (e.g., reading specialist, resource teacher, school psychologist), or perhaps to take with you for discussion at a teacher's workshop. Consider the limitations of the reproduction process you will use. Keep in mind that most copy machines—whether Xerox-type or phototransparency—do not copy color charts or colored progress lines well.

GETTING YOUR KIDS IN ON THE ACT:
TEACHING CHILDREN TO CHART

Children—just like adults—delight in seeing and hearing about themselves. From an early age, a youngster seeks his reflection in the mirror, tries to find himself in treasured family snapshots, and listens intently to his parents' stories about himself. All of these are important channels through which the child discovers and reinforces his own identity. A chart—like a mirror, like a photograph, like a story—can tell a child a lot about himself. It shows him what he "looks like" in terms of certain crucial kinds of performance.

Children often hear people say things about them that are not really clear. For instance, an adult may complain, "Why don't you stop making trouble?" Or something else like: "Shape up." "Do better in school." "Improve your reading." "Learn to pay attention." By pinpointing a precise behavior, and by showing the child a picture of how his performance in respect to this behavior looks on a chart, an adult—you, as a teacher—can show the child directly and explicitly what you are talking about.

Getting children to participate in the data collection itself is a natural first step in teaching children about charts. Show them that observing and noting their behavior is not part of a hidden agenda, the secret occupation or preoccupation of an adult whose intentions are outside the child's immediate understanding. You can reveal and share your data collection activities with the child, and get him to participate in the actual data-collection process, through such activities as countoons. After all, who could care more about his success or failure than the child himself?

Charting is not the exclusive province of adults. Children can chart, too, and sometimes children are more eager to chart than their grown-up counterparts. Maybe *you* approached the chart with misgivings based upon some of your own past experiences. But, remember that a child needn't have any of these hang-ups; and it is likely that the chart will have none of these negative associations for him. If you present it in a positive way, as a natural part of the learning environment, and as an exciting tool that he can master and use to learn about himself, charting is likely to be an interesting and challenging activity.

Any child who understands the progression of numbers, and the concept of up and down, can understand a chart. Teachers have taught second graders, and even first graders, to chart their own behavior. Try it as a part of introducing yourself and your classroom at the beginning of the year. Tell the children that charting is one way in which you will all look at yourselves, and watch yourselves grow—as students and as teacher—over the coming months. Allow each child to keep a chart about himself.

Simple programs have been created to teach children charting basics. But you can develop your own approach. You might begin by teaching the youngsters how the numbers on the chart go up. Show them what a day line is, and how to find one. Show them how to locate the number lines; how to find the *1*, the *2*, the *9*, the *3*, and so on. Let them make a data plot; teach them how to connect the dots; and then, how to read the line that shows them by its upward and downward movement when their behavior is improving and when it is not.

Children have learned how to use even semilog charts with expertise. They are capable of understanding more than you think about the proportional nature of growth that the chart demonstrates. After all, we can all see that our own learning goes through different stages. We feel differently about our performance as it becomes better. After the initial excitement—those first giant steps when we could finally hit the ball or make a song come out of a piano, or get our bike down the street without falling over—the gains we make seem smaller, more rapid, and less important. Children can relate these feelings to the semilog scale, which starts out big, and then grows smaller. It gets much harder to make that initial kind of change once you get going.

Teaching children to chart helps teachers. The most obvious advantage being that if the child can do one chart, you are freed to do more. But teaching children charting is important in other ways. It strengthens math skills, and visual-spatial skill development. And, beyond the concrete level, charting helps children gain a sense of importance and pride in themselves. But even more important, it gives them a greater understanding of who they are—not just as others define them, but in terms that they can define themselves. Ultimately, each child must take responsibility for himself and for his own actions, rather than always looking to others for ideas, direction, and support. Children need to be able to rely upon themselves and upon their own assessments of themselves as individuals. They must be able to sort out what is most important in view of their own unique strengths and capabilities.

Self-management is a sadly neglected element in the educational system. There is an underlying assumption that a child will gain these skills, but we often fail to make the concerted efforts in this area that we make regarding other instructional objectives. The chart is a concrete basis from which this kind of instruction may originate. The lessons a child can learn about himself through charting may have an impact on his future that goes far beyond the specific performance area that his chart is designed to look at.

SECTION 4
PUTTING YOUR DATA TO WORK FOR YOU

Introduction

The goal throughout this book has been to demonstrate the use of data as an exciting as well as functional tool in the classroom decision making process. Putting your data into action is a complex skill preceded by a number of smaller, less complex steps.

— Identifying a meaningful and practical behavioral pinpoint that adequately describes performance areas of interest.
— Determining how to measure behavioral pinpoints, realizing the importance of consistency in your data taking efforts (For example, recognize the importance of examining the same behavior over time, and don't change your definition from day to day. Examine carefully whether you have expressed consistency in the limits placed on performance; for instance, the number of opportunities to perform or the amount of time given for performance. Such consistency will prevent many later data interpretation problems.)
— Developing practical ideas for actually collecting the data you need in your classroom. (Remember that, in your first data collection efforts, it's wise to "start small.")
— Choosing a chart form best suited to displaying data collected on a regular basis, that is, frequently or as close to daily as possible. (Make sure that you keep your chart current. If you are letting data accumulate, you may be wasting valuable time. When you finally do get around to the actual charting of several days' data, you may have allowed unfavorable performance trends to creep in, trends you might have been able to reverse had you been charting each day.)

Beyond simply putting data on the chart, you must be able to use and make the most of what you see. This final section is designed to help you do just that. The techniques presented do not require tremendous mathematical ability to comprehend and use. They are straightforward and practical procedures to help you determine what the data you have collected are saying to you.

Overcoming Performance Anxiety

Chapter 11
Analyzing Line Chart Data

If you begin to experience free-floating anxiety at the mere mention of the words "data analysis," you are among friends. A lot of people get fidgety when they hear phrases like "regression line," "growth trend," "central tendency," or "rate of increase." That's what keeps statisticians in business. But you don't have to be good at statistics—you don't even have to want to be good at them—to understand and use this chapter. The statistics and mathematical proofs underlying the techniques presented here have already been well documented in the literature. What is left is a set of simple procedures based on even simpler rules that will help you determine what your data are telling you about pupil performance, and prompt you to make timely, accurate data decisions.

STATIC DATA PROCEDURES

Teachers frequently describe data in terms of characteristics such as range, mean, median, and mode. These are *static* analysis techniques, for they involve choosing, from among the entire collection of data points, one or two single scores which seem in some way representative of performance as a whole.

Determining the Data Range

Technically speaking, range refers to the difference between the highest and the lowest data entries on the chart. Suppose, for instance, you wish to determine the range for data whose highest score is 15 and whose lowest score is 11. Subtracting 11 from 15, you arrive at an answer of 4. This is considered a fairly low figure, indicating a relatively stable set of data points. If, however, the data are like those plotted in Figure 11:1, the range (82 − 36 = 46) is considered indicative of a much broader, more variable, less stable set of data points. And indeed, looking at the data in this figure, it is obvious that the spread is substantial.

To determine the range for data charted on semilog paper, the figures are divided, rather than subtracted, in keeping with the logarithmic basis of the chart. For instance, where the lowest data point is 11 and the highest 15, *divide* 15 by 11 to arrive at a range statement:

$$15/11 = 1.364 \text{ or } 1.36$$

This range is expressed as a "times 1.36" (i.e., the lowest number, 11, must be multiplied by 1.36 to obtain the highest number, 15). Since equal ratios (15/11 is a ratio statement) occupy equal spaces on a semilog graph, a *x 1.36* range occupies the same amount of vertical space on the chart no matter where the data actually lie.

For practical purposes, whether using arithmetic or semilog charts, people often find it more descriptive to talk about range in terms of the upper and lower limits of the data, rather than in terms of the single figure just described. For instance, they might say instead that the range was from 11 to 15. Regardless of the method adopted for reporting, however, range is only a rough, approximate indicator of data tendencies. It has two major drawbacks as a data descriptor. First, it does not reveal how the data are distributed between the extremes. Second, its value is extremely sensitive to end point fluctuations. For example, the addition or elimination of a single extreme can drastically change the value of the range, and the chance of obtaining these extreme values will be greater the more data you have.

FIGURE 11:1. *Data Range.*

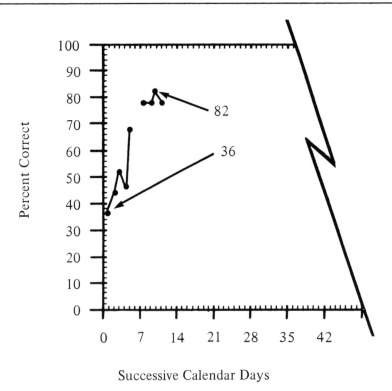

Successive Calendar Days

Determining the Data Mean

This measure of the central tendency of the data is sometimes called the "arithmetic average." It is calculated by finding the sum of the individual data values and dividing that figure by the total number of data points, arriving at an "average score." Figure 11:2 illustrates the mean figured on a set of data scores (i.e., 36, 44, 52, 46, 68, 78, 78, 83, 78) plotted on an arithmetic chart.

The mean is one of the simplest and most frequently computed data descriptors, but it is so sensitive to scores at the extreme ends of the data distribution that a mean figure alone may not give an accurate picture of the idiosyncracies of a particular child's performance. It is possible, for instance, that a child's performance may lie, for the most part, around a particular score; but a single day's extremely erratic performance pulls the mean up or down considerably, obscuring the predominant performance pattern.

FIGURE 11:2. *Data Mean.*

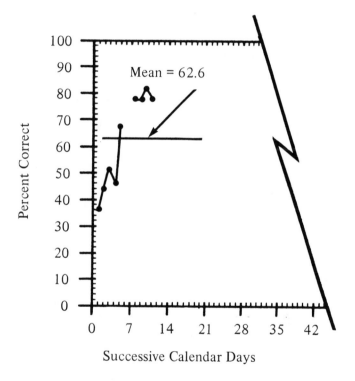

Determining the Data Median

The median is a measure of central tendency describing the precise *midpoint* of the data. Half the data lie at or above the median, and half at or below. The median for uncharted scores is figured by rank ordering the scores, then counting from either top or bottom, to decide which score is middlemost. For example, the scores 8, 2, 7, 3 and 5 are placed in rank order: 2, 3, 5, 7, 8. The score 5 is the median, since it divides the distribution so that there are two scores above it and below it. If there are an even number of scores, the median is halfway between the two middle values (e.g., for scores 2, 2, 3, 4, 5, 7, 7, 8, the median is the average of the two middle scores, 5 and 4). There are exceptions to this way of figuring medians, for instance, with a set of figures like this: 2, 3, 5, 5, 6, 7, 8. The value of the middle score is 5, but there are three scores above 5, and two scores below 5. In such a case, a median formula must be applied. This formula and its explanation can be found in any statistics text or manual.

If your data are already on the chart, there is no need to pull all the figures off the chart and rearrange them in rank or order. Simply count from lowest to highest data entries and identify the middlemost score (Figure 11:3).

Although medians are not affected by extreme scores on the ends of the data as are the measures of central tendency thus far described, they are still limited in that they do not reflect overall growth patterns.

FIGURE 11:3. *Data Median.*

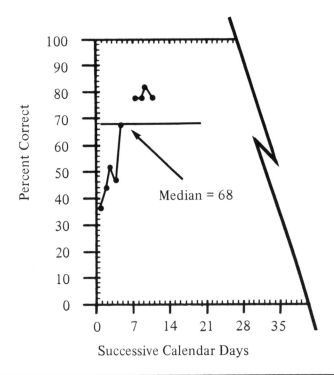

Determining the Data Mode

The mode is that value which appears most frequently in the data distribution. In a set of data such as those plotted in Figure 11:4, the mode is determined by rank ordering the scores (36, 44, 46, 52, 68, 78, 78, 78, 83), counting the frequency with which each score appears, and deciding which is most prevalent. In this case, the score 78 percent, which appears three times, is the mode. Of all central tendency measures, mode is probably the least commonly applied to individual performance data.

FIGURE 11:4. *Data Mode.*

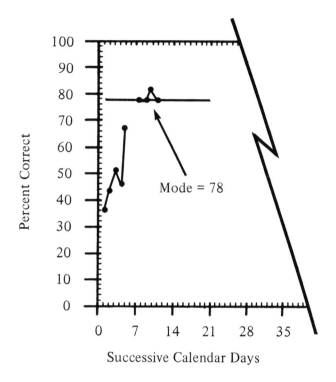

DYNAMIC ANALYSIS PROCEDURES

While static tendencies (e.g., range, mean, median, and mode) are frequently the most common, if not the sole descriptors, applied to child performance data, they may be somewhat misleading. Consider a hypothetical example: the scores of two third graders, Tanya and Celeste. Suppose that the range, median, and mode of these two sets of data are calculated, and the results are found to be identical. If these statistics alone were sufficient evidence, we would assume that the program under way is equally successful for both children. It is immediately obvious when the data are charted, however, that the children are clearly moving in two opposite directions (Figures 11:5a-b). The performance of one is increasing, and that of the other is steadily going downhill.

In this instance, and in many others, static descriptors are not enough. What is needed is an analysis procedure which answers questions about the *dynamic* characteristics of the data. How is performance changing: In what direction is the change; and what is its magnitude?

FIGURE 11:5a. *Where Static Measures Aren't Enough: Tanya's Data.*

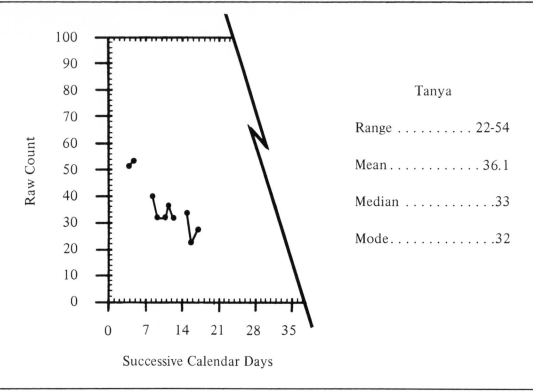

Tanya

Range 22-54

Mean 36.1

Median 33

Mode 32

FIGURE 11:5b. *Where Static Measures Aren't Enough: Celeste's Data.*

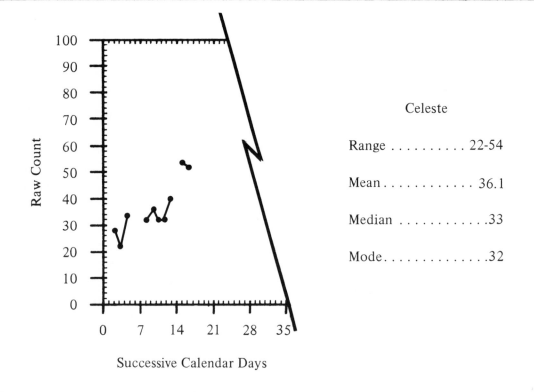

Celeste

Range 22-54

Mean 36.1

Median 33

Mode 32

Looking for a line of Progress

We have already discussed the fact that line charts are so called because the data points on them can be connected to form a line which moves upward and downward on the chart as performance increases and decreases. The overall direction these lines take, and the steepness of the incline or decline they show, has a special meaning when we think about performance growth. Some lines describe behavior increase, or an upward trend. Some show behavior to be traveling in a downward, or decreasing, direction. Others show data to be moving neither upward nor downward, but traveling a flat path. When data progress in the third manner, the child is neither improving, nor is he failing. He is simply said to be showing no growth at all, or maintaining the same level of performance over time. (See Figure 11:6.)

FIGURE 11:6. *Line Charts Show Direction of Performance Change.*

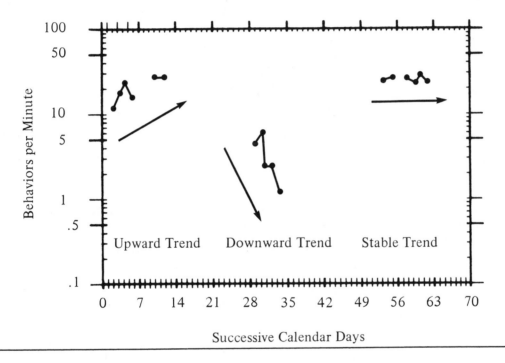

Eyeballing the chart. Often we can determine the direction performance is taking by a process called "eyeballing." We run our eye over the data and try to envision an imaginary line that is passing through it. Although there are many more sophisticated ways to analyze data trends, this rather primitive technique is one teachers frequently use, making changes when the line of progress appears to be veering from the desired course, and then quickly eyeballing the chart again to make sure that the resulting data changes appear to be headed in the right direction.

It is important to realize, however, that data are not always easily deciphered. Sometimes pattern discriminations are quite fine, and at other times, data patterns are misleading. One day's sudden drop or rise in performance may startle a teacher into implementing new and unnecessary procedures. More often, data simply "linger on." Things seem to be going well, progress is still in the right direction, so no change is made and pupil and teacher waste valuable time. Sometimes subtle, undesirable trends creep in and establish themselves before the teacher is aware of them.

These kinds of miscalculations and "surprises" can be avoided. There are ways to make data analysis more accurate and, at the same time, to give teachers the ability not only to describe performance in terms of what is happening *now*, but also what is likely to happen *in the future*.

Drawing a best fit line. A first step toward achieving improved descriptive and predictive ability requires drawing what is commonly called a "line of progress" through the data. Where, before, you tried to *imagine* a line approximating the general direction of performance change, the line of progress is one which is actually *drawn* through a set of data points, to describe behavior increase or decrease over the course of the program. This line is frequently also called a "best fit" line, because it represents the line that is "best" in "fitting" the data at hand.

The best fit line serves a function similar to that of the line of regression used in statistics to describe a scatter of data points representative of performance. If you recall any statistics training, calculating the line of regression is a fairly complex exercise. Luckily, a technique for determining the best fit line has been developed which allows you to project a line of progress without resorting to mathematical formulas or calculations.

This technique, called the Split-Middle method. has been developed, tested and refined using data charted on thousands of individual Standard Behavior Charts (White 1971; Koenig 1972). These are semilog charts, but the technique can be applied to data on arithmetic graphs as well. The operation is extremely simple: You merely locate key points on the data you have collected and connect these points with a straight line. It produces results well within the range of accuracy necessary for classroom decision making.

In order to demonstrate this method, we have taken a collection of data points off their chart and applied the following steps:

1. *Collect and chart your data.* The more data you have, the more likely is the resulting line to represent progress in general. If you want to extend the line to predict future progress, it is usually recommended that you have at least five to seven days of data. Make sure the same program plan is maintained throughout the period for which the line will be drawn. Figure 11:7a contains ten data plots. In Figures 11:7b-g, the lines which usually connect the data have been left out for ease of interpretation.

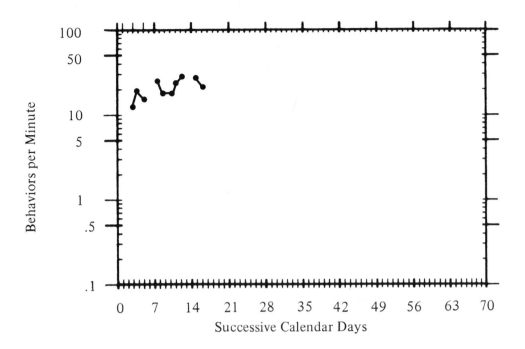

2. *Divide the data to be summarized into two equal parts.* Draw a line which divides data plots into two equal parts. If there are an equal number of data points, draw the dividing line halfway between the two middle points. If there are an odd number of data points, the dividing line will fall directly on one of the data points. In Figure 11:7b, an equal number of data plots fall on each side of the line.

3. *Find the mid-date of each half.* This point represents the day on which the middle sample for each half of the data was taken. If there are an equal number of data points on each side, the mid-date falls halfway between two plots. If there are an odd number of data days in each half, however, the mid-dates will fall directly on data points, as in Figure 11:8c. Draw a vertical line in each half which passes through the mid-date point.

4. *Find the mid-plot of each half.* The mid-plot is the middlemost, or median score for each data half. Follow standard procedures for finding a median on the chart and draw a horizontal line through the data, indicating the mid-plot for each side (Figure 11:7d).

5. *Find the intersection of the mid-plot and mid-date for each half.* Mid-date and mid-plot lines will intersect on each side of the data. Indicate these points of intersection lightly with pencil (Figure 11:7e).

6. *Draw a line through the data, connecting points of intersection.* Using a straight-edge, draw a faint line, connecting the intersection points found in the previous step (Figure 11:7f). You're almost there!

7. *Adjust the line to make a best fit.* Move the line up or down until the same number of plots fall on or above it as fall on or below it. In other words, draw a new line parallel to the one you have just drawn. This adjusted line is the true best fit line (Figure 11:7g).

FIGURE 11:7b. *Best Fit Line: Dividing Data into Two Equal Parts.*

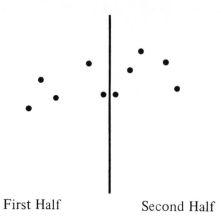

First Half   Second Half

FIGURE 11:7c. *Best Fit Line: Finding the Mid-Date for Each Half.*

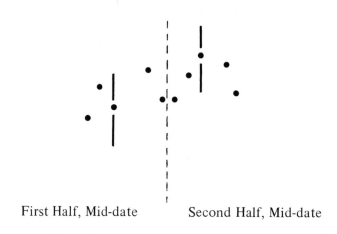

First Half, Mid-date   Second Half, Mid-date

FIGURE 11:7d. *Best Fit Line: Finding the Mid-Plot for Each Half.*

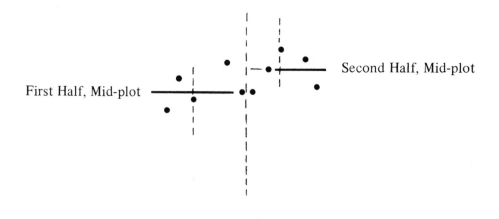

First Half, Mid-plot   Second Half, Mid-plot

FIGURE 11:7e. *Best Fit Line: Noting Intersections of Mid-Dates / Mid-Plots.*

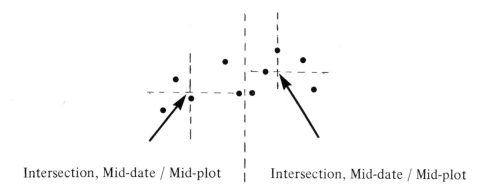

Intersection, Mid-date / Mid-plot   Intersection, Mid-date / Mid-plot

FIGURE 11:7f. *Best Fit Line: Connecting the Points of Intersection.*

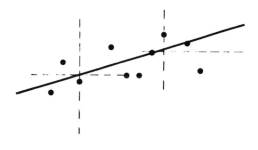

FIGURE 11:7g. *Best Fit Line: Adjusting the Line for Best Fit.*

Line adjusted so half of plots fall on, or above, or below.

Describing the Magnitude of the Line of Progress

Once a line of progress is drawn through the data, you are able to visualize with considerable accuracy whether a given performance is increasing or decreasing. You can also identify whether progress trends are steep or minimal and, by comparing one chart with another or one data phase with another, determine gross similarities and differences. Looking at the best fit lines drawn through the data for three different program phases in Figure 11:8, for instance, you can say with some certainty that the child's performance in Phase Three of the program is considerably improved over Phases 1 and 2. The program alterations in Phase 3 appear to be very effective.

FIGURE 11:8. *Eyeballing Best Fit Lines.*

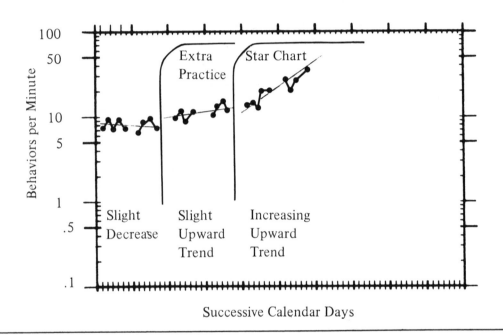

Even more information about the line of progress is available if you perform one additional operation called "trend analysis." This is a simple calculation which may be done for data on either arithmetic or semilog charts. It results in a numerical value describing the magnitude of the line of progress. The value is called a "slope," and it represents how much the child must change his performance on the average each day in order to stay on the line of progress which presently appears on the chart. On an *arithmetic chart*, the slope value tells you what *amount*, on the average, a child must add to or subtract from his performance each day in order to maintain the line of progress drawn through his data. The slope value for a *semilog chart* represents the average *percent* of change which must be made to continue in the direction indicated by this trend line.

Calculating slope on an arithmetic chart. The value of the line of progress on an arithmetic chart is figured quickly and easily by performing what we like to call the "triangle trick." Anyone who's taken a rudimentary geometry course will realize immediately that the trick relies upon the formula for finding the slope of the hypotenuse of a right triangle. Figure 11:9a contains such a triangle. The hypotenuse (marked "C") is the long side of the triangle, opposite the right angle. Its slope value is determined using the formula:

$$slope\ of\ C = A/B$$

That is, the value of side *A* is divided by the value of *B* to arrive at the value of the slope of *C*.

In order to use the triangle trick to calculate the value of a line of progress, begin by drawing a line of progress through the data on a chart (Figure 11:10b). Now turn that line into the hypotenuse of a right triangle. This is done in Figure 11:10c, where a vertical line *A* is dropped from the top of the line of progress until it meets line *B*, drawn horizontally from the other end of the best fit line. Where *A* and *B* meet, a right angle is formed, and the line of progress (*C*) becomes the hypotenuse opposite that angle.

FIGURE 11:9a. *Calculating Slope on an Arithmetic Chart: Right Triangle.*

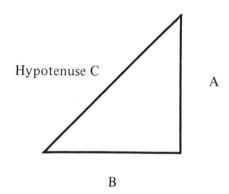

Apply the formula *A/B = slope of C* to find the value of the line of progress:

1. Determine the value of *A* by counting the number of grid marks along this vertical line (14), and multiplying by the unit value of each mark (in the case of this chart, 2). The resulting value (14 x 2) is 28.
2. Determine the value of *B* by counting the number of grid marks along that horizontal side of the triangle. Since these marks represent days, each is worth 1. The total value of this segment, then, is 14.
3. When *A* (28) is divided by *B* (14), the value of *C* is found. Here, that value is *2*.

What this value means, is that the child must add (on the average) two correct responses a day to stay on the line of progress drawn on the chart. If he does not increase performance by this much, his progress will slow down and fall off the line. If he begins to increase the number of times he performs the pinpointed response by more than two each day, the eventual line of progress will be even steeper than it is right now.

The slope in Figure 11:9 has been calculated on a line of progress which is 14 data days long. The more days of data accumulated, the greater the general accuracy of the best fit line and slope value which describes it. But, as the number of data days increases, so does the value of the line *B*, making the arithmetic involved slightly more difficult. (You may find yourself doing long division with unwieldy numbers like 17 or 31.)

Fortunately, once the line of progress is drawn, its value is the same no matter how large or small a segment of it is used to make up the hypotenuse side of the right triangle. We used the entire line in the example in Figure 11:9, but a smaller triangle could just as easily have been drawn, as in

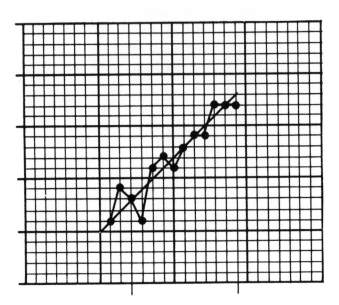

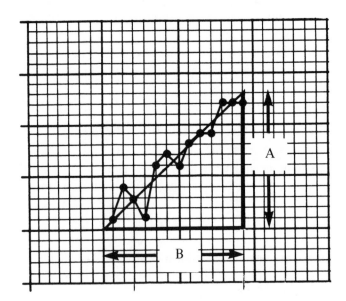

Figure 11:10. Here, the triangle is drawn on the same line of progress, but side *B* is only *ten* days long, rather than 14. The value of *A* is 20 now, rather than 28. The results, *20 divided by 10 = 2*, are the same as those for the larger triangle. Since ten is such an easy number to divide by, it is suggested that, in cases where a line of progress describes more than ten data days, a ten day segment

from the line be used to perform the triangle trick. Notice, too, that no Saturday or Sunday lines appear on this chart. The inclusion of no-chance days will affect the value of your results slightly, since you cannot assume that improvement should continue to occur even when there is no chance to perform.

FIGURE 11:10. *Calculating Slope on an Arithmetic Chart: Turning a Line of Progress into the Hypotenuse of a Right Triangle.*

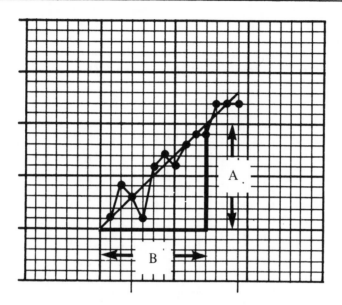

FIGURE 11:11. *Calculating Slope for a Downward Trend.*

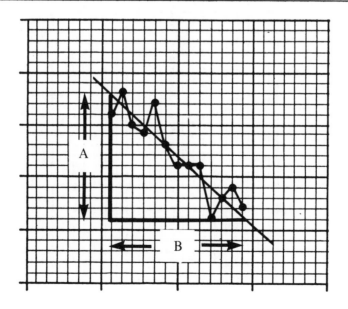

Once the line of progress is drawn, the triangle trick can be applied to any kind of data on any kind of arithmetic chart. For a slope with a downhill trend, the trick still applies, but the triangle is reversed (Figure 11:11), and the value is read as *minus*, rather than plus.

Trend values can be determined without going through the calculations required by the triangle trick if you use a "trend finder." This is a device you can create for yourself which allows you to obtain an instant slope analysis by identifying the line on the finder which most closely approximates the one on your chart and reading off the value of that line. Each trend finder must be made specifically for the chart on which it is used if slope values are to be accurate. The following directions for making a trend finder are derived from the triangle trick.

FIGURE 11.12a. *An Arithmetic Trend Finder: Darken Tenth-Day Line.*

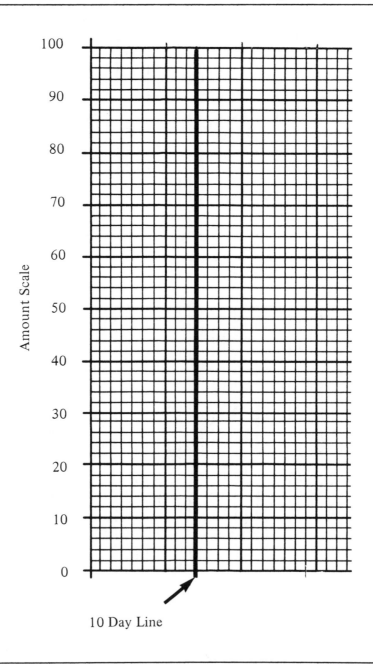

1. Starting at the point of origin (marked "0" in the figure), count off ten days and darken the tenth day line (Figure 11:12a). Ten days have been chosen to make calculation easier.
2. Count upward along this darkened line and mark the points where the line crosses the 10, 20, 30, and 40 amount lines, and so on up to 100 (Figure 11:12b). The points you have marked represent values which, when divided by 10, equal 1, 2, 3, 4, 5, and so on.
3. Draw lines radiating from the point of origin so that they connect with each mark you have made (Figure 11:12c). The radiating lines represent a series of graduated slopes whose values are +1, +2, +3, +4, and so on to +10. (If you examine the figure closely, you will see that you have drawn a series of gradually larger triangles, whose hypotenuses are the lines radiating from "0.")
4. Transfer the trend finder you have created to a piece of onion skin paper or plastic transparency, marking the slope values appropriately (Figure 11:12d).

FIGURE 11:12b. *An Arithmetic Trend Finder: Mark Off Major Intervals.*

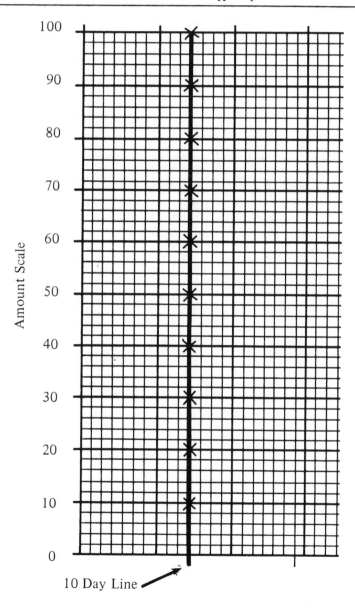

10 Day Line

FIGURE 11:12c. *An Arithmetic Trend Finder: Draw Lines to Intersect Interval Amount Marks.*

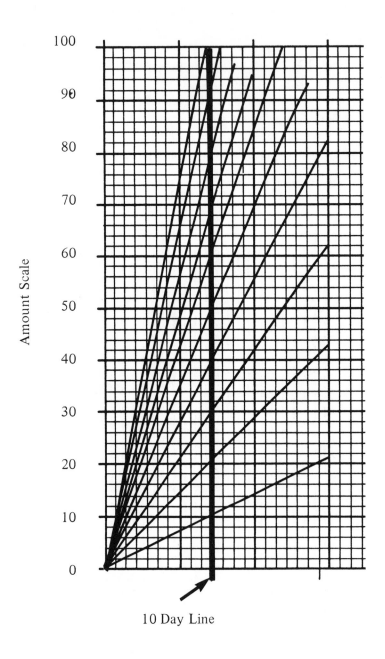

10 Day Line

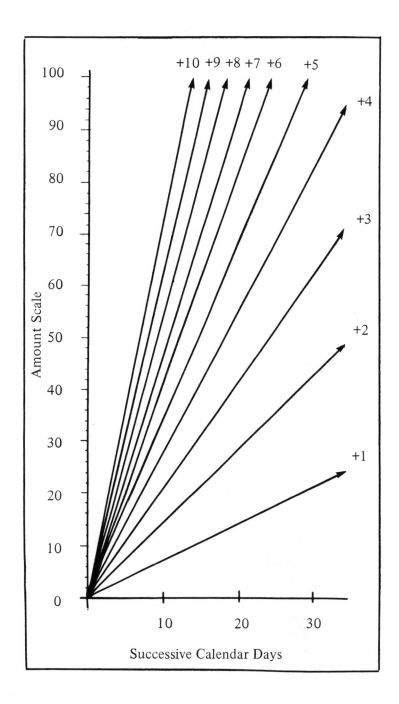

Your trend finder is now ready for use. Draw a best fit line of progress through your data. Lay the finder on the chart so that the horizontal base of the finder lines up exactly with an amount line crossing the line of progress (Figure 11:13). Determine which finder line most closely matches the line of progress you have drawn. This line's slope most approximates the slope of the line of progress on your chart. You'll notice in this example that the finder is placed on the same set of data originally used back in Figure 11:9 to describe calculating the slope value. The finder confirms that the slope of this line of progress is, indeed, +2.

Decreasing slopes are measured by inverting the finder so that each line represents a downward slope. Read the downward slope and note the results as *minus* (−) rather than *plus* (+).

FIGURE 11:13. *Using the Trend Finder to Determine Slope.*

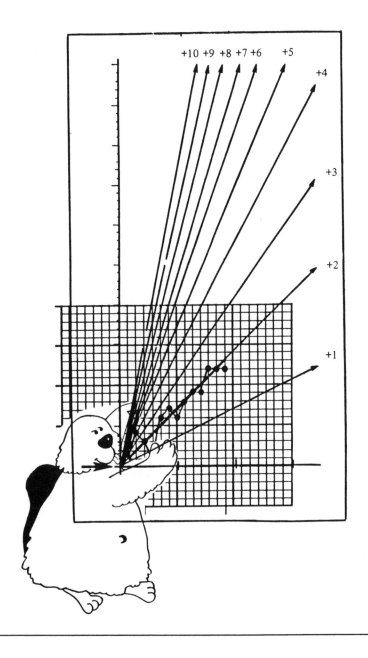

Calculating slope on a semilog chart. Figuring the slope for a line of progress is somewhat easier on a semilog chart than on an arithmetic graph. Again, the chart's special construction does most of the footwork for you. Remember that slope is interpreted slightly differently when data are plotted on semilog. On an arithmetic graph, the slope tells you the average *amount of change per day* of data traveling along the line of progress. On a semilog chart, slope describes average *percent of change*. And while it can be computed on a daily basis, slope is more commonly calculated as change *per week* in relation to semilog chart paper.

A best fit line of progress has been drawn through two weeks' data plotted on a semilog chart in Figure 11:14a. To determine the slope value of the line:

1. Start by marking off a one week segment anywhere along the line. In Figure 11:14b, this segment is marked by beginning and ending points *a* and *b*, which fall on Sunday lines.
2. Now note the *amount line* values intersected by the line of progress at points *a* and *b*. On this chart, *a* intersects the *5* amount line, and *b* intersects the *10* (Figure 11:14c).
3. Divide the value of *b* (10) by the value of *a* (5): the resulting amount is, of course, *2*.

FIGURE 11:14a. *Calculating Slope on a Semilog Chart: Best Fit Line.*

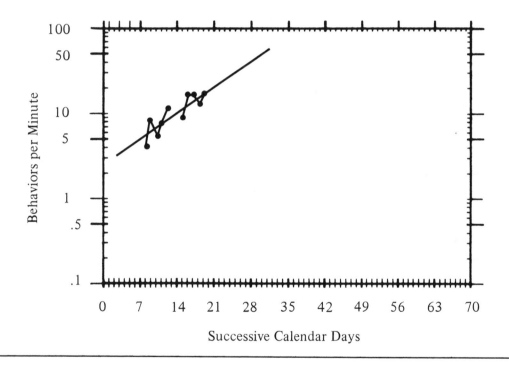

The answer, *2*, means that, on the average, the child's performance doubles (or increases by 100 percent) each week for as long as he maintains the present line of progress.

Since this is an upward trend, its slope is designated as a "times," written *x*. A decreasing slope is described as "divide by." In Figure 11:15a, the line of progress has a slope of *divide by two*, meaning that percent of performance is divided by a factor of two each week it continues along the same line. To calculate the slope on a descending line of progress, again mark off a week's segment and divide the value of *b* by *a*, as in Figure 11:15b, where *b* equals 6 and *a* equals 3. This means performance is decreasing at a *divide by 2* (i.e., by 100 percent) each week.

FIGURE 11:14b. *Calculating Slope on a Semilog Chart: Marking Off One-Week Segment* ab.

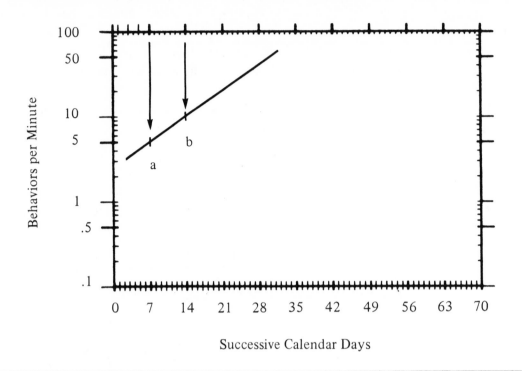

Successive Calendar Days

FIGURE 11:14c. *Calculating Slope on a Semilog Chart: Noting Amount Line Values Intersected at Points* a *and* b.

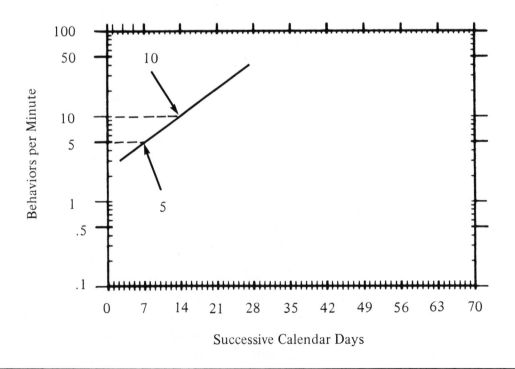

Successive Calendar Days

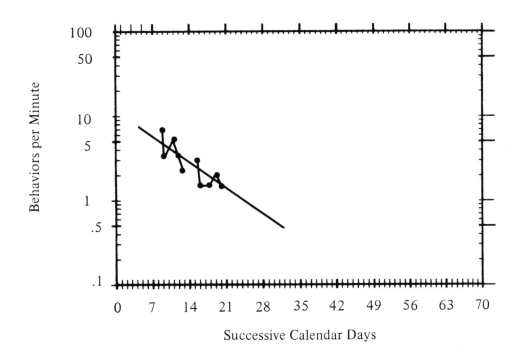

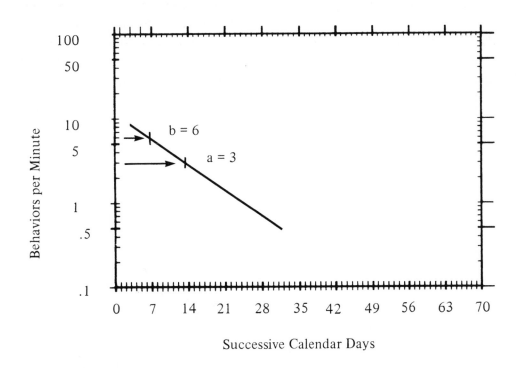

A trend finder can also be created for the semilog chart, which will make it possible to read slopes at a glance. Like the arithmetic trend finder, the semilog finder must always match the chart paper it is to be used on. Apply the following directions to your own semilog paper to devise a trend finder that works for your chart.

1. Use the 1 line on the chart as the trend finder's horizontal base. Darken the line with a pencil. Mark the first "Sunday" day line where it crosses the darkened 1 line (Figure 11:16a).

2. Continuing upward on this day line, mark the points where the Sunday line intersects each of the succeeding amount lines (2, 3, 4, 5, and so on up to 10), as in Figure 11:16b.

3. Draw radiating lines intersecting these points (Figure 11:16c). Assign number values to the radiating lines. The flat line is 1; the next line $x2$; the next line $x3$; and so on. Each line represents a successively larger slope value, ranging from 1 which is perfectly flat and shows zero percent, or no change, to a change of $x10$ per week (the steepest change on this finder). If you look closely, again you can see that you have created a succession of superimposed ever larger triangles.

4. Extend each of the radiating lines so that they travel across at least three weeks' space on the chart and reproduce the finder lines on onion skin paper or plastic transparency so that it looks like the finder in Figure 11:16d. An arrow has been added to the finder to make using it easier.

To use the trend finder, move it up and down on the chart, keeping it perfectly vertical, until the arrow on the left hand side touches the progress line to be measured. Read the slope value by noting which slope on the finder most closely matches that drawn through the data on the chart (Figure 11:17). In some instances, lines of progress will fall between the lines on this finder. In other cases, a line of progress on the chart is steeper than any which appear on the finder; and sometimes data are going downward rather than increasing. The finder in Figure 11:18 shows how additional lines may be estimated and added to the finder to show slope values of $x1.2$ (20 percent change per week), $x1.3$ (30 percent change per week), $x1.5$ (50 percent change), and so on. Figure 11:19 shows a trend finder devised to accommodate those rare instances in which performance data increase by greater than $x10$. Finally, if a decrease line of progress is the issue, simply flip over the finder and apply it to your chart, remembering that slopes which head downhill are designated by a "divide" sign.

FIGURE 11:16a. *A Semilog Trend Finder: Darken "1" Line and Locate First Sunday Line.*

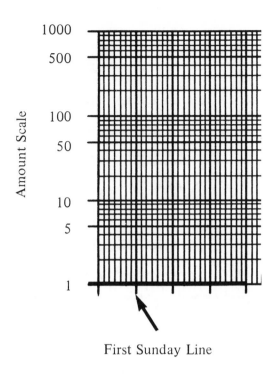

First Sunday Line

FIGURE 11:16b. *A Semilog Trend Finder: Mark Major Amount Intervals.*

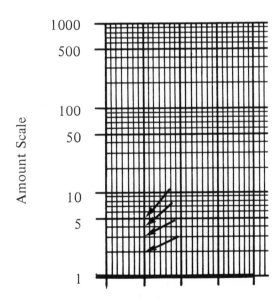

FIGURE 11:16c. *A Semilog Trend Finder: Draw Radiating Lines to Intersect Interval Amount Marks.*

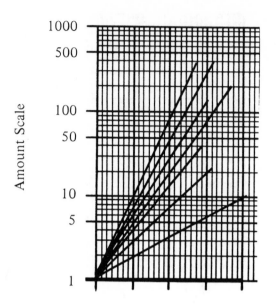

FIGURE 11:16d. *A Semilog Trend Finder: Completed.*

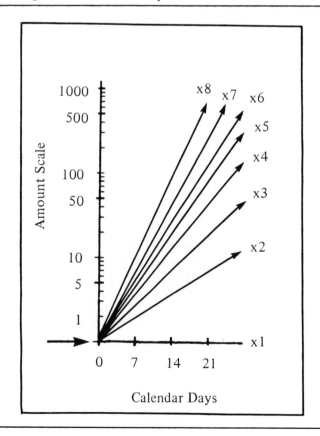

FIGURE 11:17. *Using the Trend Finder on a Semilog Chart.*

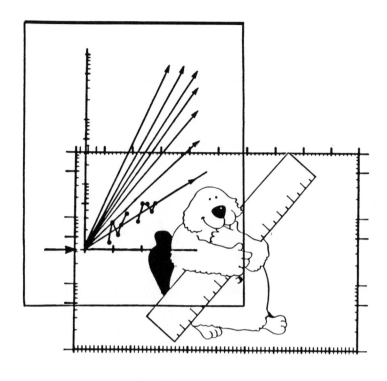

FIGURE 11:18. *Increasing Slope-Finding Options.*

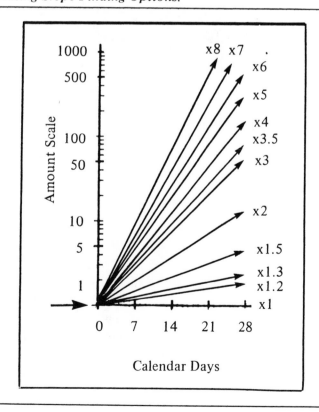

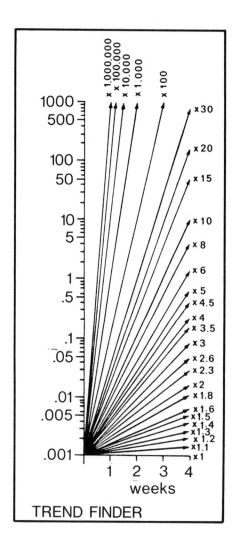

SPECIAL ANALYSIS PROBLEMS

Data analysis is like any other endeavor in life. Just when you think you have it figured out, someone throws a wrench into the works. Two data tendencies—data curve and data bounce—can make trend analysis especially difficult at times.

The Hairpin Curve

Sometimes, even though you've followed all the procedures for drawing a best fit line, the line of progress you come up with doesn't really seem to fit. Your eye tells you something's wrong. The data suddenly veer and seem headed for the top (or bottom) of the chart, as shown in Figure 11.20.

This configuration can be a special problem with the arithmetic chart. Because of the chart's equal interval construction, the only data plotted on it that will give a straight line are those in which change is at a constant amount. Any greater change will begin to form a curved line, and the greater the curve, the more difficult analysis is.

FIGURE 11:20. *The Hairpin Curve.*

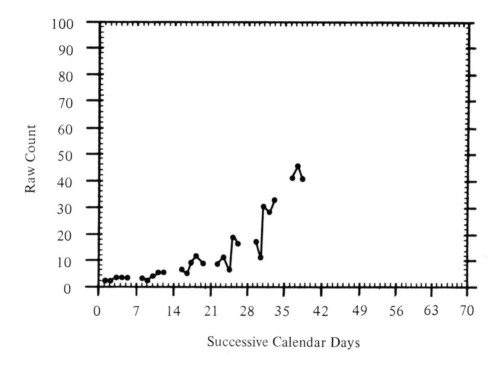

Transferring such data to a semilog chart will lessen this particular analysis problem, in the sense that a semilog chart tends to straighten data which appear curved on an arithmetic chart. In Figure 11:21, a comparison of identical data plotted on arithmetic and semilog chart forms illustrates this issue. Line A on the arithmetic chart (Figure 11:21a) represents a constant amount of increase, and thus appears as a straight line. If your data look like this on the arithmetic chart, there is nothing to worry about. In fact, going to a semilog chart with the data that increase at a constant amount, as these do, would result in a curved line, which is more difficult to describe.

As you examine lines B through E on the arithmetic chart in this figure, however, you will notice that each is successively more curved, whereas comparable lines on the semilog chart tend to be straightened. Obviously, as the lines of progress on the arithmetic chart successively demonstrate increased curvature, they become more and more difficult to analyze. You probably won't have any idea what your data will look like before they are plotted. If, once on the arithmetic chart, they present a markedly curved line, you may want to replot on a semilog chart for a more easily analyzed line of progress (Figure 11:21b).

If you are determined to keep them on an arithmetic chart, another option is to perform what is called a "piecemeal analysis." This is a technique used to segment the curved line in a manner that allows you to fit the data with two or more consecutive lines. Piecemeal analysis procedures are described in detail in Chapter 12.

FIGURE 11:21a. *Comparing Data Curves: Arithmetic Chart.*

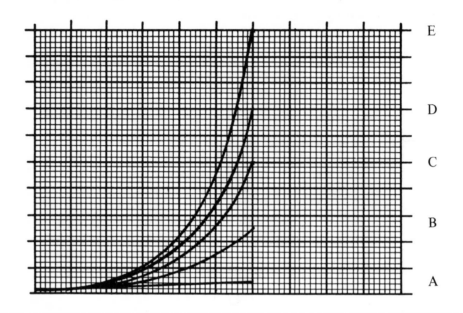

FIGURE 11:21b. *Comparing Data Curves: Semilog Chart.*

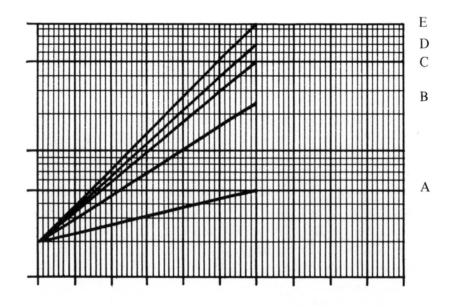

Deadly Data Bounce

Although we have made our initial examples fairly regular for the sake of illustration, data are seldom so regular that they travel smoothly along the lines of progress. In real life, individual data points may cavort considerably around the line. This type of irregularity, which tends to be more pronounced on the arithmetic chart, is called "excursion from the line," or data bounce (Figure 11:22a). When data mavericks result in bounce, it may be difficult for you to be satisfied with the picture the line of progress you have drawn provides. If you're eager to round these critters up and bring them closer to the line, remember that such data bounce can be minimized by transferring the data to a semilog chart (Figure 11:22b), where bounce tends to be compressed.

FIGURE 11:22a. *Comparing Data Bounce: Arithmetic Chart.*

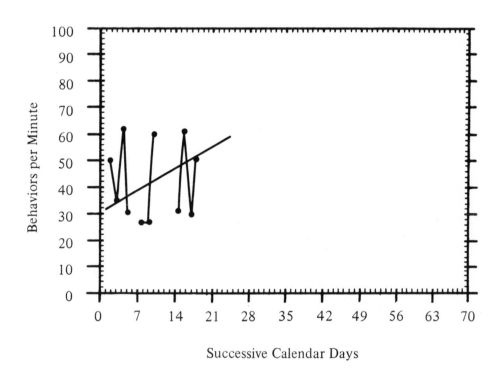

Successive Calendar Days

FIGURE 11:22b. *Comparing Data Bounce: Semilog Chart.*

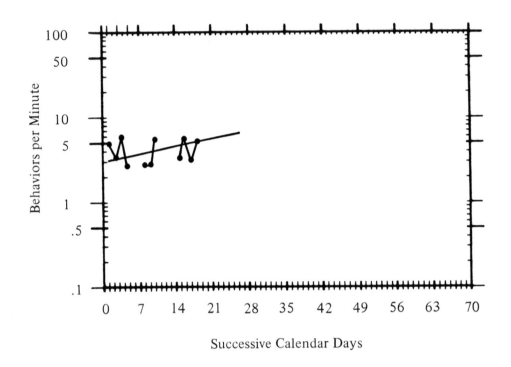

CONCLUSION

This chapter has presented two different approaches to the analysis of your data. The first, a static approach, always involves selecting, from among an entire collection of data points, one or two single scores which seem in some way representative of performance as a whole. While static procedures are useful in providing a summary statement to describe data, they overlook a critical performance dimension of which we, as teachers, must always be aware. That dimension is perform-ance change.

What happened from the time we started collecting data to the time we did our analysis? In what direction did the data move? Did the child make gains, or did he lose ground? Or, did he show little or no change in performance, simply maintaining a constant level of behavior? What is the magnitude of whatever change took place?

Data are dynamic: they move and change from day to day as you follow a child's perform-ance. If you understand the nature of this change, and can use the information to guide instructional programming, your data taking efforts take on a new meaning, becoming a functional part of your daily classroom decision making. Both static and dynamic procedures are important analysis tools. Each has its own special place in analysis efforts. Yet because each serves a different purpose, teach-ers must be clear about what type of information they are seeking and which approach will provide that information quickly and completely.

Chapter 12
Making Data Decisions

Instructional decision making often seems to demand of us something of the art of the fortune teller. All too frequently, we find ourselves sifting through the remnants of today's performance, hoping to stir up a clue for tomorrow's instructional plans. It can be a nerve-racking business, especially because—where a child's success is concerned—no teacher feels comfortable trusting to tea leaves. And, fortunately, no teacher has to.

The data analysis techniques described in the last chapter form the basis for a set of simple, straightforward decision rules. These can help teachers determine, after as little as three days' data, what the child's performance should look like, and when a program change should take place. The decision rules which have grown out of the use of classroom data have freed teachers from many of the limitations which once hampered effective decision making. They provide a set of consistent guidelines, helping teachers to define performance aims, to decide what kind of performance constitutes acceptable progress toward those aims, and to determine with immediacy and accuracy when an instructional change is necessary in order to keep performance headed in the desired direction.

Because these data decision rules have been developed by teachers who themselves use charts to monitor and guide decision making, they are constructed to be incorporated as an integral—and easy—part of daily charting activities. These decision making guidelines have transformed the behavior chart from a mere data reporting form into an *active, working, decision-making tool*.

MINIMUM PROGRESS LINES

The minimum progress line—called a minimum acceleration/deceleration line by those in *Precision and Exceptional Teaching* (Liberty, 1975)—provides an exciting technique for monitoring daily progress and, at the same time, keeping a specific performance aim in mind. The technique involves drawing a line on the chart which represents the least (minimal) amount of progress a child must make on the average to reach the goals established for him on time. In effect, the minimum progress line is a highly visual indicator of whether or not we are getting "there" from "here," that is, whether the child's ongoing progress is sufficient to allow him to reach his aim given his entering level of behavior.

Drawing a Minimum Progress Line

Using a minimum progress line to direct data decisions requires, as a first step, fixing a line on the chart which is anchored at one end to the child's present performance, and, at the other end, to the eventual performance goal. Minimum progress lines can be drawn either on arithmetic or semi-log charts, using the following sequence of steps.

— *Draw an aim star.* In order to do this, first determine the *aim amount*. This is the level of performance you eventually want the child to accomplish. Locate this level on the chart's vertical amount scale. Then decide on the *aim date*. Determine specifically when you want the child to reach the aim. Find the date on the appropriate day line of the chart. Now draw an *aim star*. Following charting conventions, draw a small aim star at the point where the aim figure and aim date lines intersect (Figure 12:1a).
— *Collect and chart three days of data.* These are the indicators of the child's present performance. Plot the data for the appropriate days according to when you start your project (Figure 12:1b). Calendar coordinate the chart to ensure that the proper number of days exist between the first three data days and the aim star (e.g., if you plan to allow 20 days to meet the aim, make sure 20 instructional days are possible).
— *Find the mid-day of the first three data days.* This will always be the second day on which data are collected. Draw a vertical line through the mid-day (Figure 12:1c).
— *Find the mid-plot of the first three data days.* This is the median point for the first three days of data. Draw a light horizontal line through the data point, extending across all three days for which data are charted, and crossing the mid-day line you have just drawn (Figure 12:1d).
— *Make a "start mark."* Draw a small circle at the point where the mid-day and mid-plot intersect. This represents a "best estimate" of the child's true performance level at the beginning of the program (Figure 12:1e).
— *Draw the minimum progress line.* Draw a straight line, connecting the start mark with the aim star. This line represents how rapidly the child must increase his performance on a day-to-day basis in order to move from his initial performance level to the aim within the allotted amount of time. Data must travel on or above this line in order for performance criteria to be met (Figure 12:1f).

Figure 12:2 shows a minimum progress line for a decrease target. The steps followed are the same as those just outlined, with a resulting line that heads downward from the start mark.

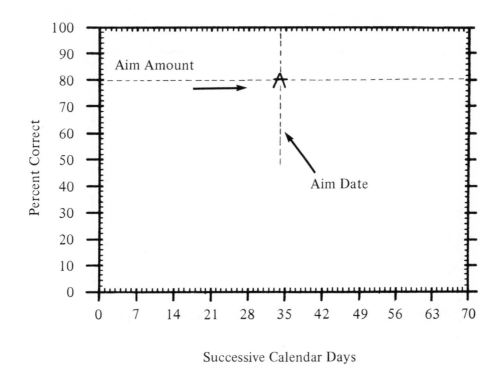

Successive Calendar Days

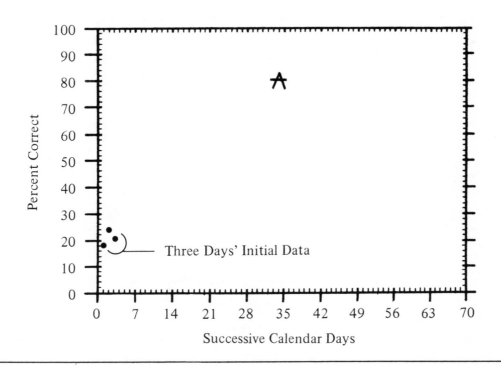

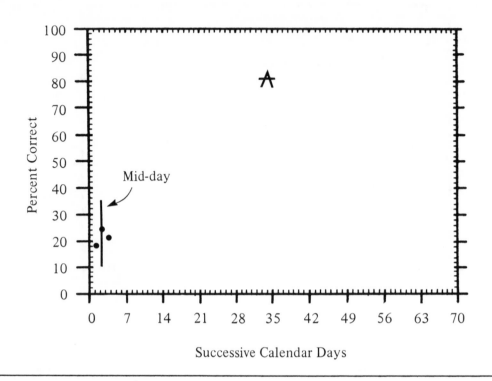

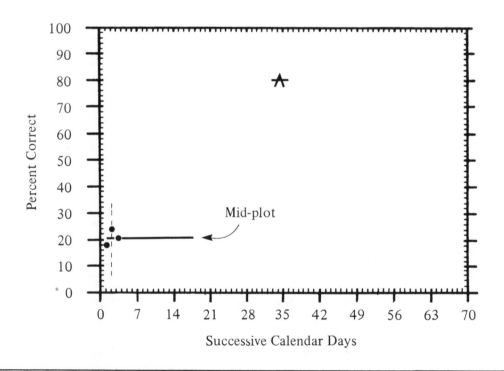

FIGURE 12:1e. *A Minimum Progress Line: Making a Start Mark.*

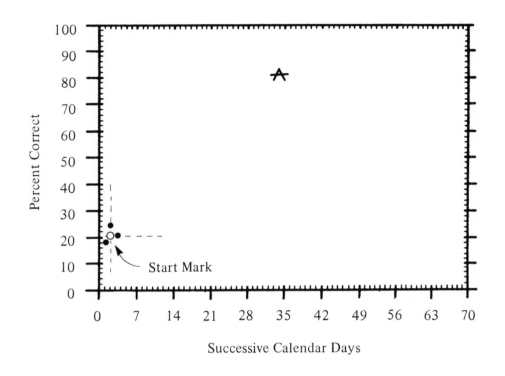

FIGURE 12:1f. *A Minimum Progress Line: Completed.*

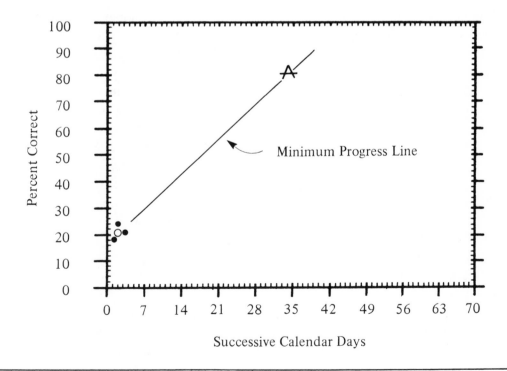

FIGURE 12:2. *Minimum Progress Line for a Decrease Target.*

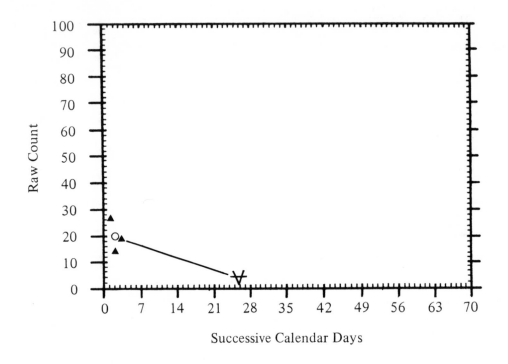

Using the Minimum Progress Line as a Basis for Making Program Changes

The minimum progress line's purpose is to act as a strong visual reminder of when a program change must be made. The first three days of data indicate where the child *is*; the aim star indicates where we want the child to *be*; and the steepness of the line indicates *what kind of change* he must maintain over time in order to get where we want him to go.

Although it is not strictly necessary to figure the value of this minimum progress line in order to employ the change rules it carries with it, you should have some idea of the nature of the change requirements you are placing on the child. The steepness of the line, or its *slope*, is a very important factor in the child's performance and in how we set aims and make program changes. If the line's slope is very steep, the child may never be able to get where we want him: He simply doesn't have the performance momentum. Say, for instance, that you have performed a trend analysis on the minimum progress line you have just established on data charted on semilog paper, and find that the line's slope is $x1.6$ (approximately 60 percent change each week). You might expect that the child will have difficulty maintaining this great a change over a long period; so you make certain adjustments to keep the criteria you set up reasonable. If the line is not steep enough, on the other hand, the child may reach the aim faster than planned and waste needless time sitting in "idle" while you plan new materials.

The following suggestions for making optimal use of the minimum progress line were developed based upon the study of large quantities of rate data, charted on the semilog Standard Behavior Chart. The hints collected here can be applied, however, to the measure and chart form of your choice.

- *Set a data rule.* Establish a rule that tells you when performance has veered from the minimum progress line sufficiently to warrant a program change. Many teachers make a change after three days' data fail to match the minimum progress line (e.g., three data points in a row fall below the line, when the goal is to increase performance; three days in a row are above the line, when the aim is to decrease the level of behavior). Others make a change sooner, after two consecutive days' performance fail to meet minimum progress criteria. The three day criterion for change was established on the basis of data charted on semilog paper (Liberty 1975), but this change criterion also seems well suited to data plotted on arithmetic charts. The important thing is to set a firm and consistent rule for change that will eliminate the "maybe-I'll-wait-one-more-day" syndrome which is so tempting to us all. Otherwise, you may find yourself trapped in a pattern in which the child hits and misses repeatedly—three days off, two on, three off, one on—until you're not using your data as much as you're letting it take you for a ride.
- *Watch what happens.* Chart data as frequently as possible (daily, if you can). Watch the data points carefully to determine whether or not the child's performance continues to meet the minimum progress line. Whenever the data values fail to meet the expectancies of the progress line for the number of days specified in your data rules, it is time to make a change (Figure 12:3a).

FIGURE 12.3a. *Data Decision Rules: Program to Increase Behavior Needs Change.*

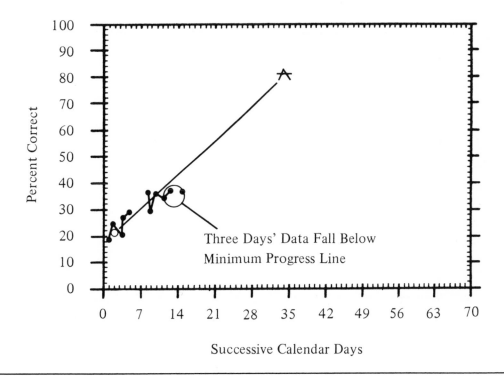

- *Initiate a change.* Try to make program changes as soon as possible after you notice that data have fallen off the line for the number of days specified in your data rule.

Make sure that the new plan will have a chance to get off the ground. If a weekend or holiday is coming up immediately after the change is to be initiated, remember that you'll have to reintroduce the plan upon the child's return, and then start your data watch anew. As a result of this break, you may lose valuable time and the program loses valuable impetus.

Take weekends and holidays into account as you contemplate a plan change. Try to allow yourself a few days to collect data on the new plan before a break appears. Watching the chart carefully should give you some advance notice of when problems may be expected to arise, with time to formulate new plans and materials before they are actually needed.

Having decided to put a new plan into effect, draw a heavy vertical (phase change) line on the chart just before the date for the new plan. This will help by making the distinction between the old and new plans readily visible.

— *Reassess the minimum progress line.* A new minimum progress line must now be drawn. This may be done using the old aim date or choosing a new one. In either case, a new start mark is established using the last three data days before the change (Figure 12:3b).

FIGURE 12:3b. *Data Decision Rules: Setting a New Start Mark.*

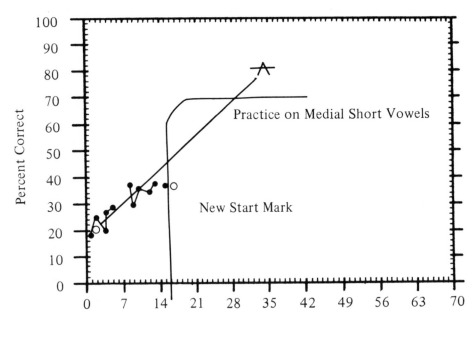

Successive Calendar Days

–294–

You may feel you have no choice but to keep the original aim date. When you draw a new minimum progress line (joining the new start mark with the old aim date) under these circumstances, however, note that the resulting line is now steeper than before (Figure 12:3c). This means that the child has to progress even more rapidly than he was asked to in the first place (a tough requirement if he was having trouble meeting the demands of the old minimum progress line). Sometimes a child is able to meet such a challenge, although his performance may drop off slightly at first.

Depending on the task involved and the child, teachers often prefer to *reset the aim date*, allowing the child more time to reach the aim. The aim date is reset by drawing a new minimum progress line, beginning at the new start mark and running parallel to the old line, until it intersects the preestablished aim amount line (Figure 12:3d).

Avoid, if possible, changing the aim amount. If you have chosen the aim carefully as a result of its value in helping the child gain increased skill proficiency, you will defeat your purpose by lowering the established performance level requirements.

— *Maintain the data watch.* When you have drawn the new minimum progress line, implement the new program plan and continue as before, keeping the plan in effect as long as the child meets or exceeds minimum progress expectation. Change the program whenever the data fall below the line or above, in the case of decreasing data for three consecutive days.

FIGURE 12:3c. *Data Decision Rules: New Minimum Progress Line, Keeping the Original Aim.*

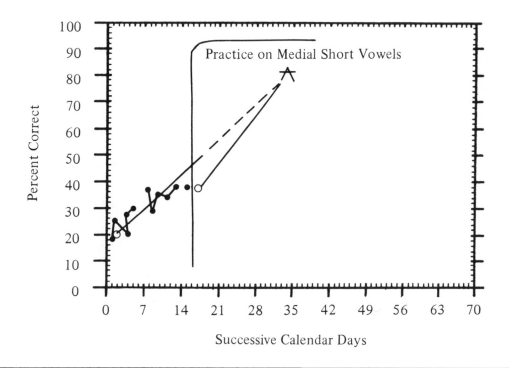

FIGURE 12.3d. *Data Decision Rules: Drawing a New Minimum Progress Line, Using a New Aim Date.*

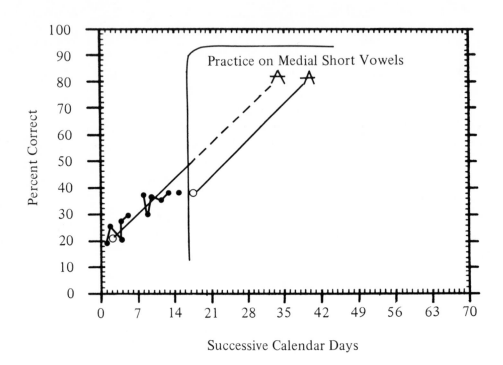

Seeing the Minimum Progress Line in Action

Using the minimum progress line is one way to keep a visual fix on how closely a child is following performance expectations. It is especially popular because it is quick, easy, and results in a simple, concrete representation of the line of progress a child must maintain in order to meet criteria. In this way, the line serves as a kind of compass which keeps our sights directed along the desired course of travel.

The following represent only a few examples of the ways in which teachers have established and used a minimum progress line to direct decision making.

Example 1. Brenda Lyons teaches a class of third graders at Valhalla Elementary. She decides to give five of her youngsters who are experiencing reading problems extra instruction based around word families. During a ten-minute period following the end of the reading group each day, Brenda looks at the children's progress using one-minute samples on randomized word lists containing representatives of the word family under instruction. All the children in the classroom are busy with independent seat work, as Ms. Lyons calls each of the five youngsters up to her desk individually and listens to them read the word list aloud for the one minute.

Brenda designs and dittos a set of reading probes. Each of the four different sheets in a probe set contains all the words selected to represent the word family, arranged so that there are 100 words per sheet, in random order. As a result of the randomization, and the fact that there are so many words per page, the children are never able to memorize a sheet, although they may read from it more than once over the two-week period they spend in instruction. She has pinpointed *say words* as the critical movement cycle for these samples, which are timed using a stopwatch, and counted with a single-bank hand tally counter as the child reads from the list. She underlines any words the child misses on a follow along sheet.

Ms. Lyons takes rate data and charts it on individual semilog charts for each child immediately at the end of the sample. She has specified an aim amount which is the same for each youngster—100 words per minute—based upon the advice of a resource person who visited the school and gave a short workshop on dealing with reading difficulties. Because Brenda has several word families to cover during the semester, she decides that she can allow a maximum of two weeks to meet criterion on each family. She uses the aim rate (100 per minute) in conjunction with the aim date (two weeks from the beginning of instruction), and three days' initial data to arrive at an individual minimum progress line for each youngster. Brenda draws these lines on each child's chart. She establishes a three-day data rule; that is, if rates fall below the minimum progress line three days in a row, a program change is made. (She sets no error criterion, again on the advice of the resource person who has told her she can disregard errors for the time being, on the theory that they will decrease automatically when the youngster is able to read 100 words per minute correctly.)

Brenda soon discovers that the youngsters really enjoy the opportunity to "show their stuff" on an individual basis for even a brief period each day. As children often come up to her desk for various reasons throughout the day, these youngsters do not feel singled out unfairly. They realize that this is a time when the teacher can pay attention just to them, and they look forward to the samples each day. In addition, Brenda finds the set-up is extremely easy to organize and carry out, and the probe sheets are something that she can use repeatedly from day to day, and from year to year.

Two of the five charts Ms. Lyons keeps are shown in Figure 12:4. Of these, one (Figure 12:4a) follows the performance of a youngster who stays on or above the minimum progress line consistently, meeting projected aims in each instance at or before the aim date. Ms. Lyons introduces a new word family each time the aim is met. In the course of nine weeks, this youngster masters five word families.

Figure 12:4b traces the performance of a second child whose rates fall below the minimum progress line before he is able to meet his aim. When performance is observed below the line three consecutive days, Brenda initiates a new program and resets the minimum progress line on the chart. Using a new start mark based upon the three days' data below the line, Ms. Lyons draws a new minimum progress line parallel to the old line and slightly below it. She extends the aim date, rather than readjusting the aim amount requirements. (Although she could have continued to use the original aim date, making the progress line steeper, Brenda feels that it would be unfair to place such requirements on children who are already experiencing difficulties meeting the aim defined.) Two plan changes are required before this youngster reaches his aim (Phases 2 and 3 on the chart: Showing the child how many words he reads correctly at the end of each sample; and Allowing him to choose and play a game when a specified rate is met). When the aim is achieved at the end of Phase 3, Ms. Lyons presents a new word family.

Brenda feels strongly that using the minimum progress decision strategy gives a sense of direction to her program planning. In addition, it allows her to give individual attention to children in the classroom who really need it. Finally, the minimum progress line permits her a concrete view of

FIGURE 12:4a. *Using Minimum Progress Lines on "Read Word Families" Data: Meeting Dynamic Aims.*

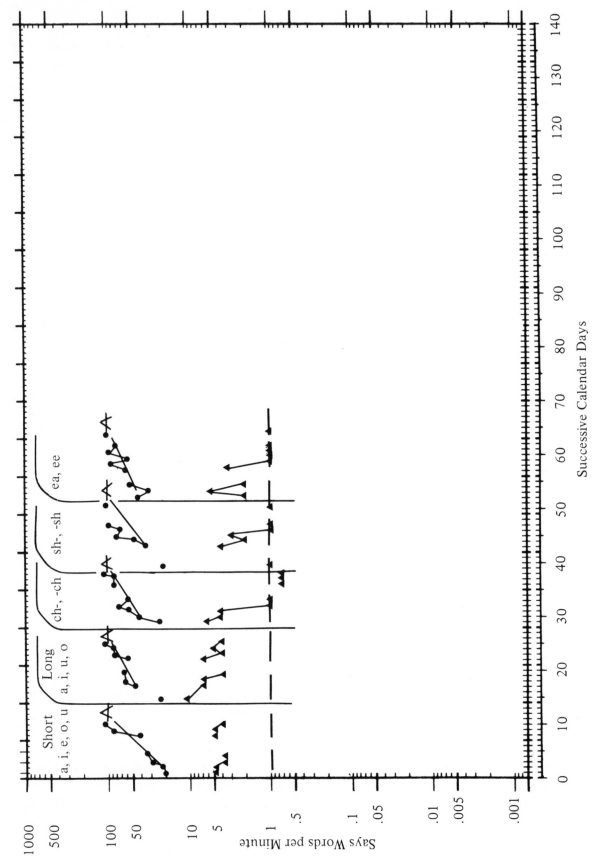

FIGURE 12:4b. *Using Minimum Progress Lines on "Read Word Families" Data: Where Program Changes Must Be Made.*

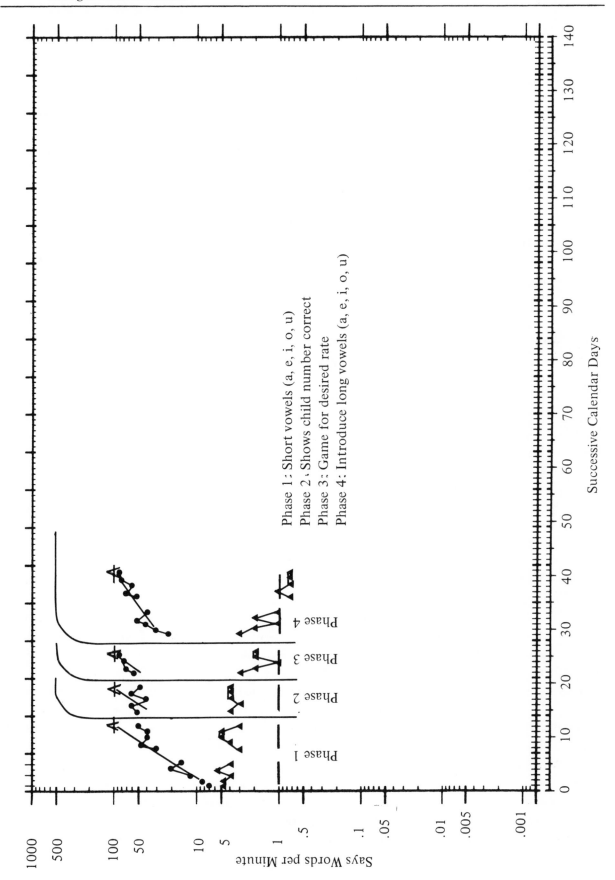

each child's progress and confirms that the extra time she spends with each youngster has positive results. In fact, she is so excited about the strategy that she has extended it to arithmetic instruction, where she now teaches "fact families" in much the same fashion.

Example 2. Heidi is a youngster in a secondary special education classroom whose frequent raucous outbursts of inappropriate laughter disrupt classroom activities so much that the teacher, Mrs. Macklin, is almost beside herself. Mrs. Macklin has tried every on-the-spot remedy she knows and nothing seems to work. She's sent Heidi out of the room, kept her late after school, sent home notes to her parents—all to no avail. Finally, she seeks advice from the instructor of an in-service course she's taking in classroom measurement. The course instructor encourages Mrs. Macklin to begin by changing nothing. She'll keep her strategies exactly the same as they are now, but take data on the number of times Heidi bursts into her raucous laugh (raw score). She's to think of this as simply an initial recording period, where she has some time to catch her breath, while she determines, with data, the exact magnitude of the problem. After counting and charting Heidi's behavior for a few days, Mrs. Macklin will decide on the introduction of an intervention.

Mrs. Macklin takes three days' initial data, counting all laughs that occur during the morning hours, between 9:00 and noon. The teacher charts her raw score data on an arithmetic graph. She sets an aim at zero, three weeks from the time she begins collecting data, and, using a start mark from the first three data days, draws a minimum progress line on the chart. Mrs. Macklin figures that she can just last this three week period: She'll continue to use the tactics she's applied all along during this time. If the data follow the minimum performance line till they reach zero, she'll be in good shape. In the meantime, she'll work to come up with a solution to what she has come to refer to as "the laugh that kills," in case things don't improve.

Mrs. Macklin's data (Figure 12:5) show that the problem is every bit as bad as her nerves have been telling her. After only three days' counting with the minimum progress line in place, Heidi's data plots jump above the line for three days in a row. According to the decision rule Mrs. Macklin adopts—three days' data off the line indicate need for a plan change—it's time for a new plan to be put into action. Mrs. Macklin is ready with one.

During this six day "period of grace," when she knew she wasn't supposed to put any new plans into effect, she has had an opportunity to *really* look at Heidi instead of her own frantic efforts, and she has chanced upon a wonderful discovery. It's become obvious to Mrs. Macklin that Heidi doesn't have any friends to speak of in the classroom, although she seems eager to get attention—albeit in a bizarre and unwelcome manner. Heidi especially seems to want attention from the boy who sits next to her during the morning period. When Heidi's teacher realizes that a new plan is called for, she resolves to use this information.

Mrs. Macklin calls a conference with Heidi and her aisle-mate, Tony, explaining to Heidi that no one in the room appreciates the "big laugh," and that in order to make friends, she'd better begin to "cool it." Tony is quick to agree. Mrs. Macklin asks Heidi's classmate if he'll help Heidi change her behavior, and make some new friends. She tells him that she wants him to make a mark on a tally sheet that she'll tape to his desk, each time Heidi bursts out laughing. Heidi can see the tally sheet from her own desk, and it is obvious to her every time the object of her affections makes a mark. Mrs. Macklin continues to keep her own count on a wrist counter, to confirm the accuracy of Tony's count.

Mrs. Macklin resets the line of progress. She draws a phase change line on the chart to mark the beginning of the new plan. Her brainstorm program has an immediate and marvelous effect. The count drops rapidly, and in two weeks' time, is at zero every day (Figure 12:5).

Example 3. Geoffrey, a ten-year old in a resource classroom, is working to improve his problem solving abilities on stories containing simple addition facts. His teacher, Alice Chinn, gives him

FIGURE 12:5. *Using Change Rules with a Decrease Target.*

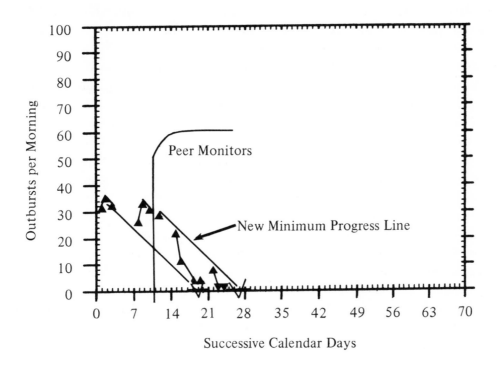

FIGURE 12:6. *Resetting a Minimum Progress Line, Continuing to Use the Same Aim.*

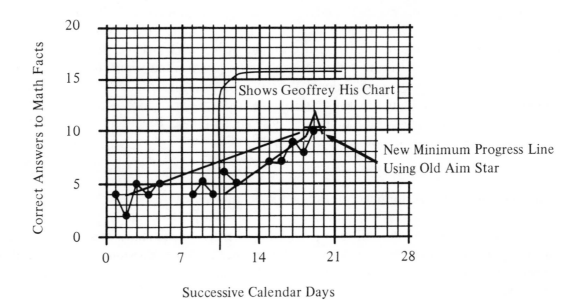

ten different problems each day, all essentially the same in the basic requirements they present, and all demanding that Geoffrey *write numeral answers* to demonstrate the solutions he arrives at. Miss Chinn always counts and gives feedback about correct and error responses to Geoff every day at the end of the math period. Later in the day she charts his raw score performance on an arithmetic chart. She decides to allow Geoffrey three weeks to meet his aim of ten correct, drawing a minimum progress line by connecting the start mark with the aim star.

After the first few days of data, which bounce somewhat around the line, Geoffrey's teacher observes that the plots topple off the line altogether, falling below minimum progress requirements for three days in a row during the second week. She makes a program change. Miss Chinn reasons that it would be beneficial for Geoffrey to chart his own performance each day: It's a perfect opportunity for him to gain more experience with numbers in a new and different way. She readjusts the minimum progress line before initiating the change, but keeps the aim date the same, since Geoffrey has deficits in a lot of areas, and really needs to move through all the material as quickly as he can. The resulting minimum progress line is, of course, slightly steeper than the original (Figure 12:6).

Geoffrey is excited at the idea of keeping his own chart, and eager to stay on the line of progress. This is the first time he's really had performance expectations defined for him in this way, and he can see exactly what he has to do to get where he wants to go. Each day he asks for help determining how many more problems correct he'll need the next day in order to stay on the line. Within only a week's time, he has reached criterion and is ready for problems containing a new set of facts. Miss Chinn feels that spending the time to help Geoffrey make one dot on the chart each day is a minimal expenditure which reaps maximum results.

ALTERNATIVES FOR SETTING MINIMUM PROGRESS LINES

The minimum progress line, as we have described it so far, is a *dynamic* projection of hoped for performance which is based upon two pieces of *static* information; that is, the start mark— a summary of initial performance, found in the mid-day/mid-plot score of the first three data days— and the aim mark, also a single point on the chart. These are connected to form a line which is a picture of how the child's ultimate performance should move and change over time.

There is another way to arrive at a minimum progress line. Already existing performance *trends* can also be used, in combination with start marks and aim information, to determine what a minimum progress line should look like. Often, teachers prefer to incorporate such dynamic information in their performance projections, drawing trends from data collected on the pupil's own past performance, or from the performance of the child's peers.

Using the Child's Past Performance to Define Expectations

Data gathered on the child's own behavior often yields the fairest and most accurate criteria for what can be realistically expected of his present performance. The following steps will help you identify what we call "past performance trends," which can be used to describe current performance goals.

— Examine the child's records and list all previous performance trends.
— Select a trend value from that list which is representative of the child's "past" performance (e.g., a trend he is able to achieve at least 75 percent of the time [Liberty 1975; White and Haring 1980]). Use this as the slope value of the minimum progress line you

will apply to all new programs. Trends can be drawn from data representing virtually any kind of performance, but those representing performance in similar tasks using similar behaviors may relate more directly to the task for which a past performance slope is being selected.

— Enter the aim criterion on the chart. This aim level can be determined in whatever way you desire. It may include the child's past performance, performance of his peers, or performance levels suggested by recognized authorities.

— Draw a minimum progress line on the chart. Line up a trend finder so that the past performance slope you have selected intersects the start mark at one end and the performance criterion level at the other. Where the past performance slope intersects the criterion level, make an aim star. Now connect start mark and aim star in the usual manner to form a minimum progress line.

The following are two examples of the use of past performance slopes to define minimum progress criteria:

Using past performance on a similar program, similar behavior. Amanda is a seven-year old in a classroom for severely/profoundly handicapped youngsters. Her teacher, Farley Granville, has been working on a program to teach Amanda to respond to sound, by turning her head in the direction of the sound whenever she hears it. At this point Granville is trying to get Amanda to turn her head in the direction of a voice (his own) calling her name. He doesn't have a lot of data on Amanda's performance at his disposal, since she's only been in the classroom for two quarters; but he does have a chart on a program she participated in at the beginning of the year which was similar to the one she's in now. The previous chart's data are on Amanda's turning her head to the sound of a noisemaker—in this case, a noisy squeak toy that Granville would randomly position slightly behind and to the right of Amanda's head or slightly behind and to the left. Amanda maintained a +2 performance slope in this program. Granville decides to require the same progress trend on Amanda's new program. He marks the eventual aim amount—90 percent, which he knows is a possible and necessary goal—across the top of the chart, then lines up the past performance slope, using his trend finder with its origin on the start mark. Granville draws an aim star where the finder's trend line intersects the 90 percent aim line, then he connects start mark and aim star, forming a minimum progress line. After eight days of performance which follow the minimum progress line, more or less, Amanda's performance falls off the line, lingering for three days slightly below. Granville adds a reinforcer (a small taste of Amanda's favorite—apple juice). Using the mid-point of the last three days' data and the same slope, he sets a new minimum progress line. Amanda's data fall mostly on target with the line until she is able to reach the established goal (Figure 12:7).

Using general past performance across a broad program range. When eight-year-old Robbie enters Mr. Terkel's learning disabilities classroom, his records state he has failed to meet his first grade arithmetic program's criteria for mastery of simple addition facts, 1-10. Already an academic year behind his peers, Robbie needs program aims that will move him as quickly as possible through material, while still guaranteeing achievement of skill mastery. Mastery means different things to different teachers, and may vary considerably depending upon the individual child and the skill area in which he must perform. Mr. Terkel defines mastery using rate data, as performance at a desired rate criterion maintained for three consecutive days. These rates must be high enough to assure success when the child enters the next, more difficult skill level.

In this specific case, substantial add-fact data (collected over a period of years) indicate to Terkel that 30 correct per minute with no errors is sufficient. Mr. Terkel will leave error considerations out of the picture temporarily. If Robbie meets the correct aim of 30 per minute and errors are still a problem, attention will be focused on incorrect responses. The math program Mr. Terkel has devised consists of a series of one-minute samples, designed originally as probes. These sheets have variety enough to allow them to be used easily for ongoing assessment of Robbies progress.

To establish a minimum performance line, Terkel compiles Robbie's past rate data in *all performance areas*. He notes trends for each phase on every chart, and identifies a trend Robbie seems easily capable of demonstrating 75 percent of the time. This trend is x1.3, 30 percent increase per week. Combining this slope with three days' start mark data and an eventual aim of 30 per minute, Mr. Terkel draws a minimum progress line on the chart, and monitors Robbie's performance in respect to it (Figure 12:8).

Robbie should reach his aim in about nine weeks, but his initial performance is erratic. Usually far above minimum progress, it occasionally dips below, and finally—in the program's third and fourth weeks—takes a nosedive. Lack of motivation seems a plausible factor. Mr. Terkel decides to share performance expectations with Robbie, marking a desired rate on each math sheet with a happy face stamp. The new minimum progress line, based on the original aim date, is slightly steeper. Excited at first, Robbie works to reach or pass the happy face each day. Then he slumps again. In a third plan, Terkel pairs desired rate with a contingency. Robbie earns a connect-the-dots constellation puzzle to work if he reaches the happy face. He is delighted with the puzzle contingency, searching through his own prized stargazer's guide to identify each new constellation earned. Robbie reaches his aim only one day after the aim date. He is ready to go on to facts 11-20 when he has maintained aim rates for three consecutive days.

FIGURE 12:7. *Using Past Performance to Determine Minimum Progress (Similar Program / Similar Behavior).*

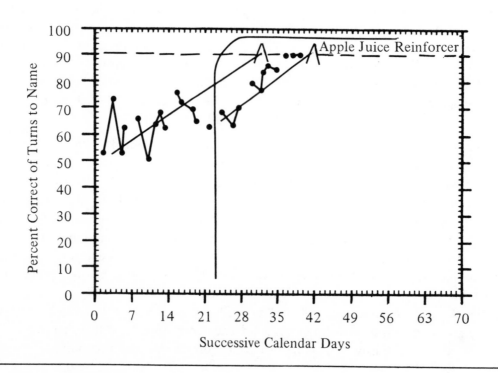

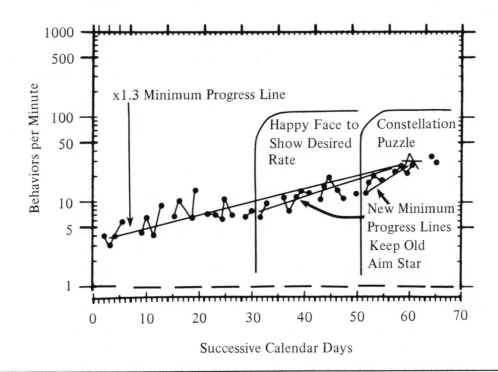

Using Peer Performance to Define Progress Expectations

Sometimes the performance of peers (youngsters in the same program, school, or other environment) on a similar or related task can be used to establish performance criteria. The performance of peers is the most difficult to relate to that of the child himself, but it offers some standards for establishing performance expectations, especially if the child is trying to catch up to peers. There are a number of ways to use peer performance in setting new performance criteria.

Static criteria (e.g., desired rates, percentage scores, raw scores, etc.), obtained using only a few days' data, or perhaps even a single score, can be used to determine an aim star which is eventually incorporated into a minimum progress line. Locate peers and sample their performance on the same tasks you will be teaching. For instance, you might ask the teacher of a classroom in which you hope to place the child, if you could observe briefly some youngsters she considers to be "good," "average," and "poor" pupils as they perform the specified tasks. Your data from this observation will define a range of performance in the classroom you visited and give you a concrete basis on which to set criterion levels. These can be entered on the chart without reference to an expected date of completion, or they can be combined with criteria for completion time and entered using an aim star. Using the resulting aim star and three days' initial data, a minimum progress line is drawn.

Another method involves using *peer performance slopes* to establish program aims. Such slopes are obtained by sampling the performance of a number of the child's peers—children in other programs in the school or in his own class. (The more peers you can observe, the better; but one or two "good" or "average" pupils will provide the necessary information.) Take data for a number of days—between five and 11 are needed for an adequate slope—to see how these youngsters perform on the task in question, and use the data trends you find to set aims for the youngster whose performance you want to improve.

These slopes may be used to define minimum progress in two different ways. The first uses slope only, with no statement of final aim amount.

1. Locate a start mark on the chart, based upon three days' initial data.
2. Enter the selected peer slope on the chart, lining up the trend finder in conjunction with the start mark and draw the slope line on the chart. This is the minimum progress line.
3. Decide how many days you desire the youngster to maintain performance on or above the minimum progress line. When criteria are met, a program change takes place. If data fall below the line for three consecutive days, a program intervention is introduced.

Peer performance slopes may also be used in conjunction with aim amount to establish a minimum progress line:

1. Locate a start mark on the chart, using three days' initial data.
2. Designate an aim amount (determined by whatever criteria desired) on chart.
3. Enter peer performance slope on the chart, lining up the trend finder in conjunction with the start mark. Where the peer performance slope crosses the aim amount line, draw an aim star. You have now established a desired aim date (intersection of peer performance slope and static aim amount) and a minimum progress line upon which performance is to stay as the child progresses toward completion of the objective.

Some say that in order for children to catch up with their peers, they have to move even faster (that is, show performance increase on a steeper slope) than the youngsters with whom they are compared. This sounds harsh, but if a child's performance is behind that of his peers, he does, in a sense, need to improve at a faster rate, if he ever hopes to overtake the other youngsters. Careful program slicing and other instructional strategies designed to meet the individual's needs may make this possible. If you use peer performance data to establish expectations, try to keep in mind what is possible and what is not possible for the child you are working with.

Using static aims and peer performance slope information. Rosemary Zapata uses a simple screening strategy with the six year olds entering her first grade classroom each year. One of the questions she wants to answer is which children can identify letters of the alphabet, printed in both capital and lower case. She has discovered in recent years that most of her youngsters exposed to the alphabet—through Sesame Street, intensive preschool experiences, and practice in kindergarten—know it upon entry. She wants to be able to single out those youngsters who require a little extra help so that the entire classroom isn't put through unnecessary instruction.

Rosemary presents a set of mixed probe sheets (capital and lower case letters randomized on sheets with 100 items each) in one-minute samples, which she times and scores for individual youngsters at different points throughout the day, when all the other children are involved in individual seat work tasks. Ms. Zapata asks each child to point to the letters on the sheet she gives him, and read the letters aloud until she says "Stop." She charts the results, scored on a follow-along sheet, at the end of the day.

Ms. Zapata takes a total of five days' data for each child. She plots these rate data on a semilog chart. Her programming decision rules involve two components. Youngsters are singled out for extra help if they perform both one-half below the class median (which turns out to be 40 correct per minute); and if they also show an overall trend of less than $x1.25$ (25% increase per week, which most of the children seem capable of maintaining).

When the children's data are displayed on a single summary chart (Figure 12:9), Rosemary sees a number of patterns emerge. Some children perform above the median rate, more than 40 per minute, with slopes of $x1.25$ or greater. These youngsters will move on to other tasks immediately at the end of the five-day screening period. Others with a high median performance seem to be growing less rapidly, some even begin to drop off slightly. As long as their median performance is at least 40, however, Ms. Zapata is unconcerned about these youngsters. She figures they are probably bored and will "pick up" as soon as they are presented with a more stimulating task. Those who really need special attention are the ones who fall below the class median and show flat or decreasing growth trends over the five-day period. Of the twenty children in Ms. Zapata's class, five show a median performance at least half below the class median. Of these, two (labeled "A" and "B" on the reach the class median performance unassisted before long. Three others (labeled "C," "D," and "E") seem to require extra assistance, as their median performance is low, accompanied by performance trends that are less than 1.25 or decelerating.

For these three youngsters, she uses the data she has gathered during the initial screening period to establish a start mark. (The start mark is placed where the first three days' data mid-plot and mid-date coincide.) The line of progress information Rosemary has acquired from her screening will now also be put to use. She could choose the lowest upward trend seen in the class as a peer performance slope for these children, but she decides that such a slope would not be adequate. While these kids are just acquiring the skill, the others have not only gained some knowledge, but have been growing better every day in relation to the level they came in at. To catch up to their peers, these three youngsters will have to go faster than the slowest child and even faster than those in the middle. For this reason, when Ms. Zapata determines the minimum progress line, she uses the highest class slope (i.e., $x1.8$). If this proves to be too steep, she can always make later adjustments. (In picking the highest slope, Ms. Zapata follows the philosophy of those educators who believe that children who are behind must go even faster than most of their peers in order to catch up. Depending on your own philosophy, you could use the median slope for the class or any other slope that showed growth.)

Using peer performance slope. The high school seniors in an honors contemporary issues class are required to meet certain speech and debate standards, defined by their teacher, Buzz Eggleston. Buzz has organized what he calls the "public forum," a one-week unit for the 20 youngsters in the class. Eggleston asks the students to split up into groups of five each for the last half hour of each class period. Every day, five discussion questions are given to each group. Although the set of questions is uniform across groups, no group ever overhears the content of any of the others' discussions. Questions are divided among a group's members so that each student gets his own "hot issue"—a topic which falls in line with class discussions and readings for the two week period preceding the forum unit. The youngsters take five minutes to organize their thoughts, and then each takes a turn within his group, stating an argument and listing supporting points concerning his question. Each student is allowed three minutes to present as many points as he is able (criteria for acceptable performance are worked out by the students themselves). The other students in the group listen, counting correct responses. These raw counts are quickly compared and charted on arithmetic paper at the end of each forum period.

Youngsters enter the class with a wide variety of self expression skills. Some can communicate well, expressing their opinions forcefully and articulately on a variety of topics and issues. Others are much less skillful. Buzz has no absolute goals for the class as a whole. He is concerned only that each student grow in terms of his own performance at class entry. He wants the good speakers to improve just as much in relation to their own initial performance as the poorer speakers. For this reason, Buzz helps the students set minimum performance lines by drawing a +1 slope line from

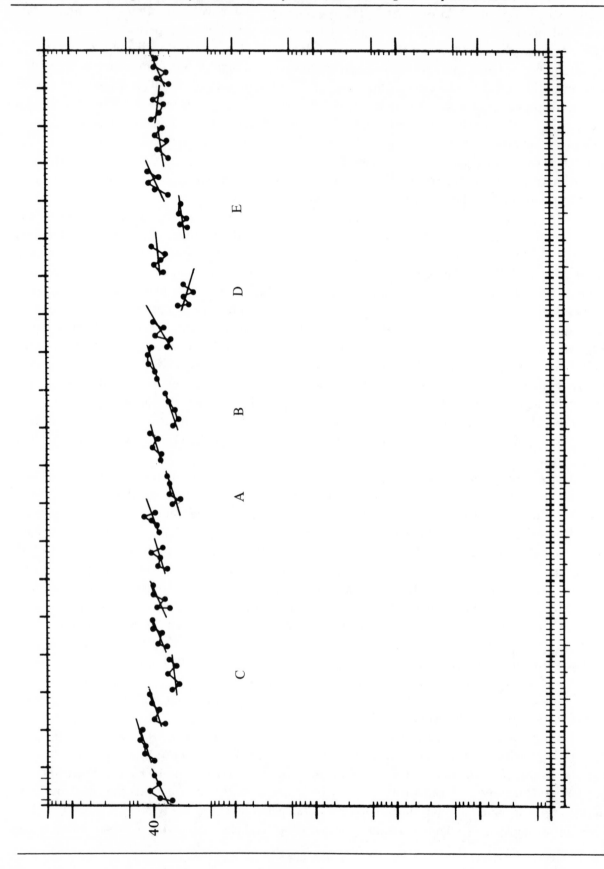

FIGURE 12:10a. *Using Peer Performance Slopes to Define Minimum Progress in Current Events: Student Needs Further Help.*

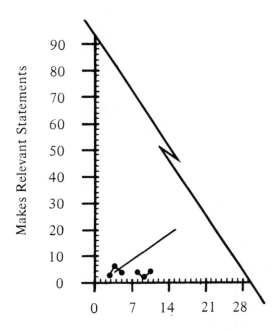

FIGURE 12:10b. *Using Peer Performance Slopes to Define Minimum Progress in Current Events: Student Exceeds Minimum Expectations; Needs Challenge.*

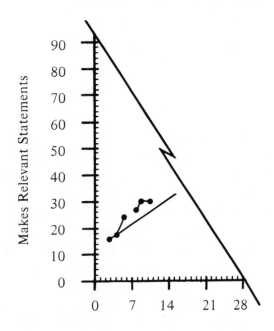

their own individual start marks across on an arithmetic chart. To stay on the indicated minimum progress line, each youngster must improve his performance by adding one new statement, on the average, each day. Performance should stay on the line for one week. Over the many years he's taught, Eggleston has found this +1 slope to be possible for virtually all pupils. At the end of the week forum period, each youngster is expected to still be on or above his minimum progress line. Students who fall below the line form a special group which continues to meet each day until performance requirements are met. Figure 12:10 shows how Buzz interprets the charts of three students in his class.

FIGURE 12:10c. *Using Peer Performance Slopes to Define Minimum Progress in Current Events: Uneven Performance Merits Further Examination.*

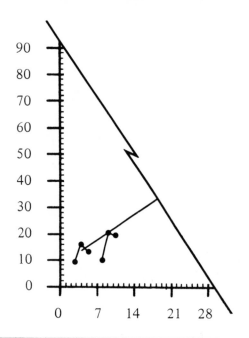

Setting a "Standard" Line of Progress

If no other trend criteria for progress can be identified, a "standard" line of progress can be applied to rate data charted on semilog paper. The concept of a standard progress line—called a standard 'celeration line in Precision and Exceptional teaching—was developed several years ago on the basis of data from several hundred programs dealing with children of all ages performing all types of skills (Liberty 1975). Research showed that of those children whose programs produced *some* progress, 50 percent were able to grow at a rate of *times* 1.33 or better in programs designated "to increase" behavior, and at a rate of *divide by* 1.46 or better in programs designated "to decrease" behavior problems. About 53 percent of the children grew at *times* 1.25 for acceleration targets, and 66 percent at *divide by* 1.25 for deceleration targets. Based upon these data, you might expect that a child would have a 50 percent chance of progressing at, at least, *times* 1.25 or *divide by* 1.25, depending upon the direction of 'celeration desired. This trend represents an average weekly increase of 25 percent.

For this reason, where no other performance trend criteria can be determined, a teacher using semilog chart paper might expect the youngster to demonstrate a progress line whose slope equals *at least* 1.25.

This means of establishing a minimum progress line is likely, of course, to be much less sensitive to the requirements of individual task and performer characteristics than any of the others presented in this chapter, and at the present, no data exist to describe its application to any measures other than rate, nor to any chart paper other than semilog.

USING PIECEMEAL ANALYSIS: WHAT TO DO
WHEN THE MINIMUM PROGRESS LINE DOESN'T QUITE WORK

The minimum progress line forces us into confrontation with our data on a daily basis. It requires that we evaluate a child's performance each day in relation to criteria for progress that we have laid out on the chart. The nicest thing about it, is that it permits us to carry out this constant weighing procedure without the necessity of performing tedious and time consuming arithmetic calculations each day to see whether the additional performance the child demonstrates is enough to keep him on track for our eventual goals. The line on the chart takes care of all the calculations; we simply watch the plots to see that they don't stray too far. If data fall off the track, we know it immediately and can initiate a rapid change *before* we and the child have suffered through days or even weeks of needless failure.

In the examples you have seen so far, the data decisions were extremely clear cut. As soon as performance fell off the projected minimum progress line, according to predetermined rules, a change was made. In most cases, the data presented have stuck fairly close to the projected progress line after these adjustments took place. The line adequately described the child's usual progress.

But, as handy as it is, there's nothing magic about the minimum progress line. It's not a magnet that draws data automatically to it. Sometimes, for a variety of reasons, a child's real performance is not adequately described by the minimum progress line you have projected. For instance, the child's actual performance may far exceed minimum progress; his data rise above the line immediately and seem to take off. It is obvious that you have somehow miscalculated in establishing a minimum progress line. Maybe the three days' start mark data weren't representative. (For some children, teachers use start marks based upon more than three days.) Maybe your aims are simply too low, or you've allowed far too much time to reach criterion. Whatever the reason, you want to capitalize on what you now see as the child's real potential. You need a different kind of decision making strategy.

Consider the data in Figure 12:11. For awhile the data are parallel to but far above the line. It is apparent the teacher was happy with these results and failed to notice when other progress trends began to set in. After 11 weeks above the minimum progress line, the youngster's performance finally fell below, necessitating a change. But maybe the change should have taken place much earlier, when it became obvious that the child's initial momentum started to slacken. Based upon his first weeks' data, the child might be expected to meet criteria much sooner, making it possible for him to go on to new material. Instead, he spends an entire 16 weeks on a single word family. Because the teacher was attending only to whether data were above minimum progress, she failed to see the pattern that began to emerge. Had the teacher been watching more closely, it might have been obvious that the upward trend of the data began to level off about five weeks into the program (Figure 12:12). Here a strategy such as piecemeal analysis (introduced in the previous chapter) might have rescued this child from what turned into a data slump.

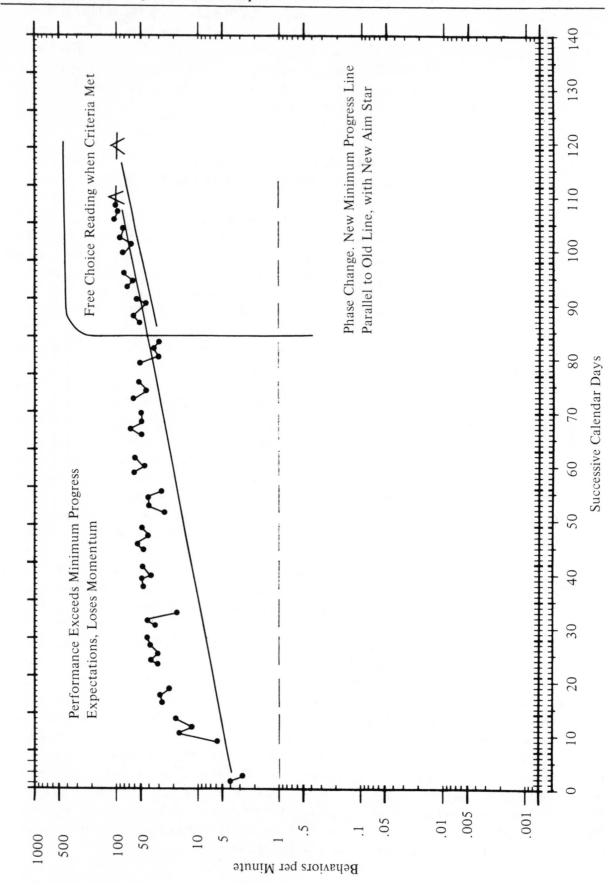

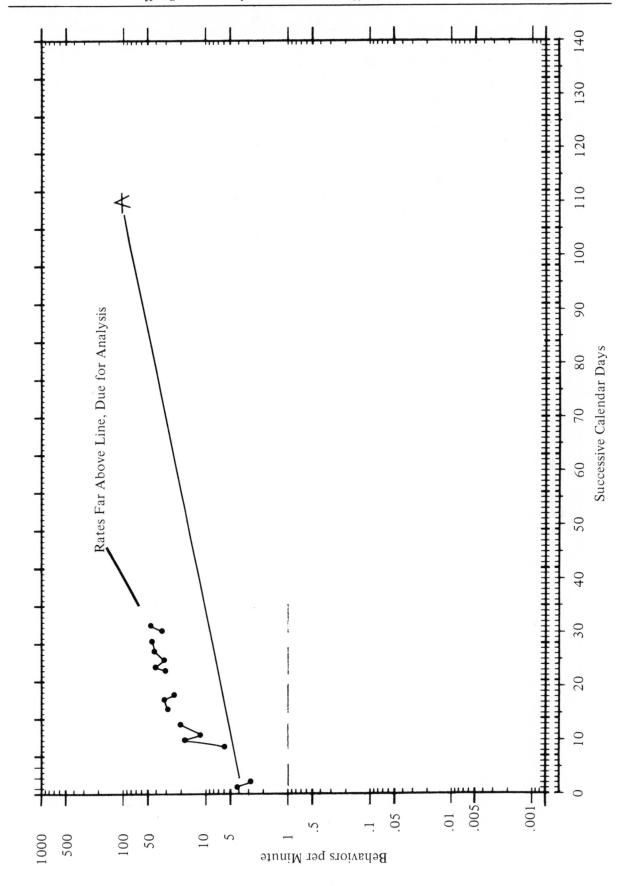

As you'll remember, piecemeal analysis involves breaking data apart to analyze smaller, more representative pieces of the total line of progress. On the basis of considerable rate data collected on Standard Behavior Charts, some educators (Liberty 1975; White and Haring 1980) have recommended analysis using segments including seven consecutive data days. Figure 12:13 demonstrates how such an analysis is performed, using the partial data from Figure 12:12.

1. Mark off the most recent seven data days on the chart. These are the only days that will concern you for the time being (Figure 12:13a).
2. Divide data plots in half (Figure 12:13b).
3. Locate the mid-data point and mid-data day of the first half of the segment (Figure 12:13c).
4. Locate the same points for the second half of the segment (Figure 12:13d).
5. Draw a line through the data, connecting intersection points on both sides (Figure 12:13e). Adjust the line to form a best fit as described in Chapter 11.
6. If the trend is flat or going in a direction opposite from that desired, an immediate change is called for.
7. If the data are still headed in the desired direction, additional analysis may be required to determine whether a subtle change is already beginning to creep in. Go back another seven days and perform an additional piecemeal analysis (Figure 12:13f). Compare the two piecemeal slopes. If the first slope on the chart is steeper than the second, the child is beginning to lose momentum and you may choose to initiate a change or continue to perform piecemeal analysis on future data every two or three days. (Always use seven data day segments, which include the most recent data plots.)

Piecemeal analysis as it has just been described is a troubleshooting technique which allows you to identify a generally undesirable trend and put a new plan into effect before it's too late. The technique may also be used to help you determine a new and more appropriate minimum progress line for a particular child.

The piecemeal analysis technique was developed with rate data plotted on semilog chart paper, but it is especially useful on data charted using arithmetic graphs. As we pointed out in the previous chapter, straight line analysis is often difficult on arithmetic charts. The piecemeal technique will allow you to zero in on and describe trends occurring when a single trend line fails to adequately describe data exhibiting considerable bounce or curve.

FIGURE 12:13a. *Comparing Seven Day Slopes to Determine Need for Change: Marking Off Most Recent Seven Days.*

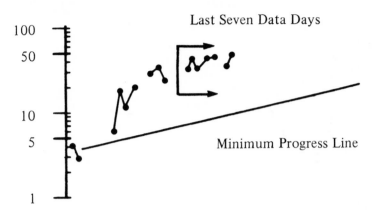

FIGURE 12:13b. *Comparing Seven Day Slopes to Determine Need for Change: Dividing Data Plots in Half.*

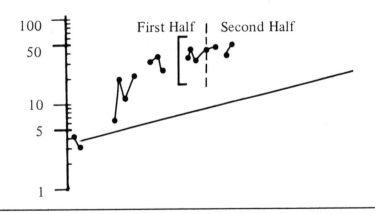

FIGURE 12:13c. *Comparing Seven Day Slopes to Determine Need for Change: Locating Mid-Date / Mid-Plot Intersection for First Half.*

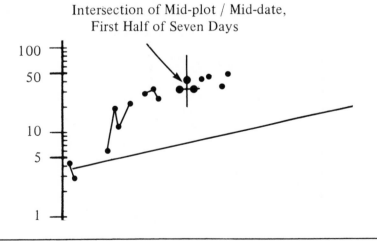

FIGURE 12:13d. *Comparing Seven Day Slopes to Determine Need for Change: Locating Mid-Date / Mid-Plot Intersection for Second Half.*

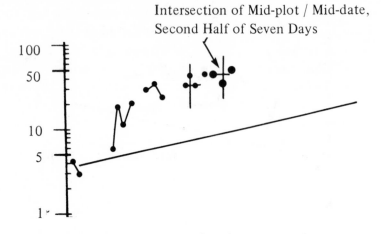

Intersection of Mid-plot / Mid-date, Second Half of Seven Days

FIGURE 12:13e. *Comparing Seven Day Slopes to Determine Need for Change: Connecting Intersection Points and Making Best Fit Adjustments.*

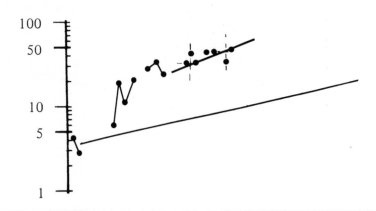

FIGURE 12:13f. *Comparing Seven Day Slopes to Determine Need for Change: Comparing Two Piecemeal Slopes to Determine Need for Change.*

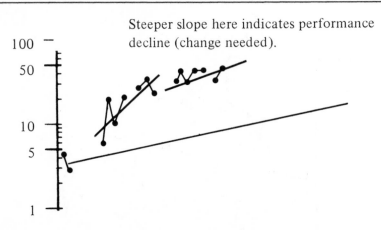

Steeper slope here indicates performance decline (change needed).

CONCLUSION

Data decision rules help turn two dimensional charts into a three dimensional picture of learning. They allow us to combine the child's performance past and present to predict his progress in the future. Charts do not dictate what we can or cannot do in the classroom; they do not prescribe methods, materials, programs, or props, a particular philosophy of education or style of teaching that we should adopt. They merely reflect, rather, the results of our instructional decisions. In teaching we are always looking for answers: What book, what approach, what consequence, what strategy *works*? Charts become our most powerful teaching tool when we use them to ask these important questions. Data speak for the child; and after all the consultants, methods, materials, books, and courses, it is still the child who must tell us what is best for him.

References

Bloom, B. S.; Engelhart, M. D.; Furst, E.; Hill, W.; and Krathwohl, D. R. 1956. *Taxonomy of educational objectives; handbook 1: cognitive domain.* New York: David McKay.

Cohen, M. A., and Gross, P. J. 1979. *The developmental resource: behavioral sequences for assessment and program planning.* New York: Grune and Stratton.

Kunzelmann, H. P. (project director). 1973. Progress report III on July 1, 1973, of State of Washington's Child Service Demonstration Programs in Seattle-Spokane-Tacoma for Precise Educational Remediation for Managers of Specific Learning Disabilities Programs. Superintendent of Public Instruction, Division of Curriculum and Instruction, Special Services Section, Olympia, Washington.

Kunzelmann, H. P. (ed.); Cohen, M. A.; Hulten, W. J.; Martin, G. L.; and Mingo, A. R. 1970. *Precision teaching: an initial training sequence.* Seattle: Special Child Publications.

Koenig, C. H., and Kunzelmann, H. P. 1980. *Classroom learning screening.* Columbus, Ohio: Charles E. Merrill.

Liberty, K. A. 1975. Decide for progress: dynamic aims and data decisions. Working paper no. 56. Seattle: University of Washington, Experimental Education Unit, Child Development and Mental Retardation Center.

Lindsley, O. R. 1971. Precision teaching in perspective: an interview with Ogden R. Lindsley. *Teaching Exceptional Children* (Spring) 3: 114-119.

Lovitt, T. C. 1977. *In spite of my resistance, I've learned from children.* Columbus, Ohio: Charles E. Merrill.

Silvaroli, N. J. 1969. *Classroom reading inventory.* Dubuque, Iowa: William C. Brown.

Strunk, W., and White, E. B. 1979. *The elements of style*, 3rd ed. New York: Macmillan.

Walls, R. T.; Werner, T. J.; Bacon, A.; and Zane, T. 1977. Behavior checklists. In *Behavioral assessment: new directions in clinical psychology*, eds. J. D. Cone and R. P. Hawkins. New York:

Brunner-Mazel.

White, O. R. 1971. The "split middle": a "quickie" method of trend estimation. Working paper no. 1. Eugene, Oregon: University of Oregon, Regional Resource Center for Handicapped Children.

White, O. R., and Haring, N. G. 1980. *Exceptional teaching*, 2nd ed. Columbus, Ohio: Charles E. Merrill.